ISDN

A Practical Guide
To Getting Up and Running

ISDN

A Practical Guide
To Getting Up and Running

Completely Updated and Expanded 2nd Edition

BY WILLIAM A. FLANAGAN

ISDN USER'S GUIDE
copyright © July 2000

Published by CMP Books
An Imprint of CMP Media Inc.
12 West 21 Street
New York, NY 10010

ISBN 1-57820-048-2

For individual orders, and for information on special discounts for quantity orders, please contact:

CMP Books
6600 Silacci Way
Gilroy, CA 95020
Tel: 800-LIBRARY or 408-848-3854
Fax: 408-848-5784
Email: telecom@rushorder.com

Distributed to the book trade in the U.S. and Canada by
Publishers Group West
1700 Fourth St., Berkeley, CA 94710

Manufactured in the United States of America

To

Dee, Alyssa, and Brendan

Contents

Preface
To the Second Edition

In a sense, the dust has settled after the battle, and ISDN is still standing. But its army has dispersed—last year the ISDN Users Group held its last meeting, at which it declared victory and dispersed.

They were right, again "in a sense." ISDN has indeed matured: it is widely available, simpler to order than five year ago, and much more competitive in price. However the same five years were the first of a phonomenon called the World Wide Web, based on the Internet. Hardly anyone saw in advance the runaway success of Internet-based services and recreation. It amazed me too. The nature of Internet usage changed almost everything in communications.

ISDN first gained a boost from early Internet users. They found 64 kbit/s dialup was super-premium access compared to the 14.4 to 33 kbit/s modems most people had in 1996. They liked it so much they stayed on-line for hours (modem users did too, but they needed the time to finish their big downloads ;-). The Internet also taught us creative use of punctuation.

Far from making telephone companies happy, this surge in their business made them scream in pain. The calls to Internet service providers were tieing up the local phone switches with calls that lasted much longer than the three minute average on which telcos based their network designs. Internet calls were choking the local exchanges, forcing switch expansions, new trunk installations, and other headaches. All those phones off-hook used much more electrical power.

Worst of all, most of the minutes of "Internet" usage generated *no* usage revenue:

- calls came from "flat rate" residential lines, no per-minute cost;

- calls went to an ISP's regular business lines, for only the monthly charge (which is why LECs asked the FCC to designate ISPs as long distance carriers—so they would pay several cents per minute in access fees).

Bummer, as we've learned to say.

Carriers reacted predictably: they tried to raise ISDN rates, but generally failed. In the late 1990s, they found another form of salvation: Digital Subscriber Line.

DSL technology, on the same copper pair as the analog phone, not only transmits Internet data much faster than modems or ISDN, it shunts that Internet traffic away from the voice switch (to a router or data switch in the central office). DSL is covered in the Fifth Edition of my *T-1 Networking* book.

Pricing residential DSL service at about the cost of an ISDN BRI line and a cheap ISP, a local exchange carrier (LEC) gives the heavy Internet user a good deal on faster access and gets him (her) off the voice switch. Everybody wins.

Except ISDN.

Remember, digital ISDN and analog voice are both part of the public switched telephone network, on the same switches. As part of moving Internet access from the switched network, LECs are hiding ISDN from residential users. No mention on web sites, no promotions, no sales efforts for ISDN to the home.

For business, ISDN is now the "fancy" voice access service, with all the features described in this book. It's ideal for large call centers. OK, you can have it for dial backup of data circuits too, because you won't be using it much (and you'll still pay every month for the line).

So the new position for ISDN is a specialized service, one among many, no longer all things to all people.

- Highly featured voice.
- Dial backup for data services like frame relay.
- But not a participant in the Internet rush, as are DSL and cable modems.

All that said, my BRI line is working just fine. The router with the built-in terminal adapter lets me run five computers on the Internet at the same time (some DSL vendors discourage or prohibit more than one). I can connect my little PBX and Fax machine too (still a bit dicey on a cable modem). ISDN isn't all things to *all* people, but it's pretty good for me.

William A. Flanagan
Box 411, Oakton, VA 22124
May 2000

Preface
To the First Edition

The Integrated Services Digital Network has taken fifteen years to move from "I Still Don't kNow" to "It Sure Delivers Now." Part of the delay was caused by a long-standing focus on the technology by the carriers, who were slow to develop a user's viewpoint. As my publisher Harry Newton put it, "If the telephone companies sold sushi the way they sold ISDN, it would be called cold dead fish."

All that is changing in the US as it has changed in the rest of the world. Western European phone companies (mostly government-owned monopolies) enforced ISDN as they converted from analog to digital switches in their central offices. In those countries, users have little choice besides ISDN. Eastern/Central Europe will do the same, as they build a modern telecommunications infrastructure, to obtain the many benefits of ISDN (particularly the greatly reduced amount of copper cable needed in the outside plant to support a given number of phone lines).

In the US the benefits are more economic. Local exchange carriers see ISDN switched circuits as the much-less-expensive re-

---

placement for many forms of labor-intensive special services, like leased lines. As data traffic continues to grow many times faster than voice traffic, the traditional ways of provisioning data circuits are becoming impossible to support. Multidrop analog leased lines, the staple of SNA networks for more than 20 years, are simply too much trouble to install and maintain.

To get away from dedicated lines, the telcos want to install ISDN lines, then let the customer decide how they will be used—POTS voice, Switched 56, access to frame relay or ATM services, etc. Setting up a new service (or activating service when a customer moves into a building) will then be done at a keyboard, perhaps by the customer himself. This will minimize labor costs. ISDN uses switches and cross-connects to set up circuits, not manual placement and connection of wire. ISDN also reduces the potential for mistakes (most T-1 outages have been caused by craftsmen working on main distribution frames as they install other lines).

Tariffs reflect the savings carriers expect to realize. When RBOCs get serious about ISDN, they make its cost very attractive compared to leased 56 kbit/s local loops for many applications. As ISDN experience accrues, expect the attractiveness of ISDN to increase. Tariffs will be adjusted to "migrate" users off leased lines.

Experience will be necessary, as will improvements in ordering procedures. As this is written, it is very difficult to get answers to even mildly technical questions from the people who take your order for ISDN service. And there are many questions to answer: configuration of a central office ISDN port may require up to 50 decisions, mostly based on the kind of customer premises equipment (CPE) installed and the services desired.

Fortunately, Bellcore and the carriers recognize the problem and are trying to simplify ordering by establishing several "profiles" for known applications and types of equipment. Applying a profile answers most of the 50 questions. In time, hardware makers will adapt their equipment to use certain profiles and eventually we hope to see "plug and play" for the most common types of equipment. This has happened in Europe, where BRI is fairly fixed.

ISDN A Practical Guide

Until then, however, you are well advised to plan ahead and be prepared. That's the purpose of this book, which is intended for the network operator who is considering or encountering ISDN for the first time. It is not detailed enough to guide equipment design. (The source materials bulk into several cubic feet of paper; major sources are in the Bibliography, Appendix B.). Nor is this book intended as the first thing to read for people new to the telecom industry. My earlier book, *T-1 Networking*, is a good starting point; two other volumes, *Frames, Packets, and Cells* and *ATM User's Guide*, offer information about data protocols and higher speed transmission systems that complement the contents of this book.

In writing about ISDN, I went through what you face: ordering and configuring lines, setting up CPE, figuring out why it didn't work, and making it function. My hope is that my experience will make yours easier.

You can share your knowledge and experience as well, in future editions of this book. Your suggestions and questions you'd like to see answered are welcome and encouraged.

William A. Flanagan
Box 411, Oakton, VA 22124
September 1995

Acknowledgements

The people at the Bell Atlantic ISDN service center endured my questions and probing for deeper answers with good cheer and patience. They helped me learn the practical aspects of installing ISDN lines and services.

Special thanks are owed to my wife and family who put up with too much because this book was started immediately after moving to a new home.

And yet again, it was publisher Harry Newton whose foresight and perception made this book possible.

Thanks also to the many, many other people at switch makers, carriers, and CPE vendors who were kind enough to share their knowledge and experiences with me. In particular, special thanks to Roy Illingsworth for commenting on the draft manuscript.

1.

Where ISDN Came From

The "digital network" in the name explains only some of its origin. "Integrated Services" are the heart of the concept, which means that many different applications can be carried by the same network. This is exactly the same goal that later drove development of ATM technology (cell-based packet switching). That is, traditional voice (plain old telephone service), video, images, and various forms of data (particularly packet switched data, which includes Local Area Networks) would be carried by the same network. To mix the different forms of traffic, all would be digitized into bit streams at speeds compatible with the central office switches.

The ISDN would be a complete solution to any telecommunications problem for any user—based on a circuit-switched infrastructure. To a certain extent, it has succeeded: there is voice, obviously, but also dial-up video conferencing and "ISDN routers" and "LAN modems" used for Internet access.

ATM, Frame Relay, and Internet Protocol later offered similar visions based on packet switching. Ironically, frame relay started as another "bearer service," in addition to voice and H.323 video, intended to use ISDN dialed connections.

Digitized Voice As the Start

"Mixed streams of digitized information" sounds today as if it started with data. In fact it started with voice.

In the mid-1960s telephone companies in large cities ran out of room for copper wire in under-street ducts. Bell Labs came up with a solution: multiplex 24 voice channels on two copper pairs, using "channel banks" *(Fig. 1-1)*. Instead of using a separate pair for each connection, channel banks made at least 12 times as much use of the wires then in place by carrying 24 connections. If the channel banks carry E&M connections, which require up to eight wires each, the two pair can replace up to 192 wires. Channel banks saved tons of copper and avoided digging up thousands of miles of streets.

Without the public realizing it, channel banks converted the transmission system inside the telephone company from analog to digital. Switches remained analog far longer—past the year 2000 in a few areas. Channel banks defined voice transmission down to the present and, at the same time, laid the foundation for ISDN.

Voice Channel Defined

To represent the sound of a voice in the form of 1s and 0s was not an easy engineering task in 1964. Semiconductors were still new, crudely simple by today's standards, yet they had to perform as well (sound as clear) as the analog phones that had been under development for almost 100 years. The solution relied on compromises in sound quality that had been made much earlier in the telephone sets themselves:

- Small physical sizes of audio components (microphones and speakers) meant they couldn't reproduce bass frequencies below 300 Hz (herz, or cycles per second). No bass also protected the user from "60 cycle hum" induced by AC power lines.

- Almost all of the information content in a voice is carried by frequencies below 4000 Hz. Therefore no special effort

Digitizing, Multiplexing Voice Save Wire

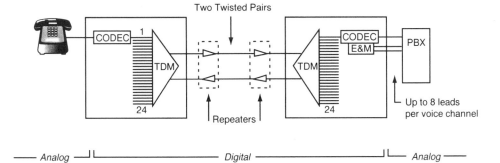

Figure 1-1

was made to reproduce voice in "high fidelity": there was no treble either.

- Listeners prefer the loudness of the received sound to be relatively constant. Thus, no provision was made to reproduce a wide dynamic range.

Therefore the new digital-voice channel bank had to deal only with the mid-range frequencies in a narrow volume range. These assumptions greatly simplified the task, while still allowing good quality reproduction of voice.

Standard Digital Encoding

Research (Nyquist's theorem) had proved that a faithful representation of an analog signal, like voice, could be made by periodically sampling (measuring) the loudness of the sound at a rate that was at least twice the highest frequency of interest in the sound. That is, measurements have to occur twice as fast per second as the highest frequency, in Hz. Measuring "often enough" ensures that you detect every change in the analog signal from positive to negative ("zero crossing") and can record it in the digital signal *(Fig. 1-2)*.

Telephones filtered out practically everything above 3500 Hz just from mechanical constraints. Loading coils in the outside plant (added to long local loops) act as filters to cut off everything above about 3000 Hz. In designing a channel bank, 4 kHz was taken as

Digital Encoding of Analog Voice

Analog samples converted to digital number; pulse code modulation

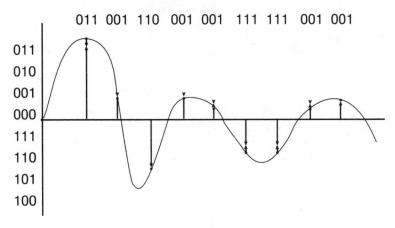

011 001 110 001 001 111 111 001 001

*Sample the voltage and represent instantaneous value with
a binary number*

Figure 1-2

the absolute top frequency. To further define the limit, low-pass electronic filters were added to channel banks, just to make sure nothing above 4000 Hz reached the measurement circuits.

A channel bank measures loudness at twice this frequency: 8,000 times per second. In an analog telephone, "loudness" is the difference in loop current from the value that flows when there is no sound input (idle loop current). Each analog measurement, called a sample, is converted by a codec (COder/DECoder, a mass-produced chip) to 8 digital bits, which can represent up to 255 values. The middle of the range represents idle loop current, which is increased and decreased to match the sound transmitted. Thus half the digital values represent positive sound pressure; half, negative pressure. The all-zeros value is not used (which ensures 1s density on the digital T-1 line).

The conversion is not linear, but optimized to work with the three compromises, particularly the one that limits loudness to a relatively narrow range. Making the conversion non-linear also saved digital bandwidth (this first voice compression achieved a 2:1 reduction in the number of bits that had to be transmitted for a given sound quality).

DS-0 Channel Defines ISDN

Eight bits per sound sample were chosen for convenience in processing: 8 is the number of bits in an alphanumeric ASCII character in digital form, and computers handle 8 bits easily. The choice of 8 bits per sample, combined with the sampling frequency of 8,000 per second, defined the Digital Signal level zero (DS-0) which we still use as the standard voice channel.

Multiplication of 8 times 8,000 gives the famous 64,000 bit/s for a channel. The T-1 rate of 1.544 Mbit/s is the sum of 24 channels (1.536 Mbit/s), plus 8,000 bit/s for framing overhead. (For complete details of voice digitization and T-1/E-1 formats, see the author's *T-1 Networking*.)

The DS-0 also forms the basis of ISDN services. It is the "Bearer channel" that carries a user's information. A digitized voice channel, the first "bearer service," occupies a DS-0 when encoded in the standard way with Pulse Code Modulation (PCM).

As mentioned, there are 24 DS-0s in a T-1 (1.544 Mbit/s).

There are 32 DS-0s (including one for synchronization and one for signaling) in a 2.048 Mbit/s E-1. When one of their DS-0s is devoted to ISDN signaling, called a Data or D channel, a T-1 or E-1 is called a Primary Rate Interface (PRI). The remaining channels carry user traffic: the bearer or B channels. Thus a PRI is 23B+D or 30B+D. The ISDN Basic Rate Interface (BRI) has two bearer channels and one D channel (2B+D).

Digital Switching

Channel banks provided digital service between central offices *(Fig. 1-3)*. But at both ends each voice channel was still analog, on individual wire pairs (with perhaps additional wires per channel for signaling). The interfaces on all voice switches were still analog in the 1960's.

Analog switches provide a wire-like connection that passes any form of signal-including the modulated direct current from a telephone or a channel bank. Analog switches didn't know about digital voice or T-1 lines.

Wouldn't it be convenient if the digital transmission line (the T-1) could be plugged directly into the switch, instead of pass-

Digital Transmission Between Analog Switches

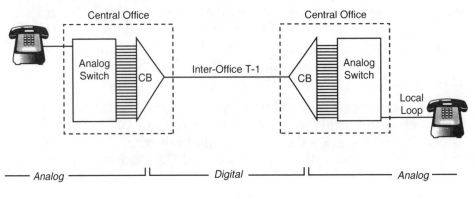

Figure 1-3

ing through the channel bank? Yes indeed, and it happened quickly as central office switches evolved from analog internals (stepping switches, crossbar switches) to digital architectures (time slot interchanger). When a T-1 line connected directly to a digital switch, the carrier saved not only the channel bank hardware, but also 23 cables and 23 (expensive) ports on the switch.

Digital switches are very different from analog switches. Digital means passing bits, not analog currents. The digital switch has to look like a repeater or amplifier rather than a wire, and it has to understand digital signaling.

When the first digital central office switch was installed, it provided what we today call switched digital service. The caller, by dialing, obtained a DS-0 connection on demand. The caller didn't realize what was happening because the phone was always connected to an analog port on a channel bank, not directly to the switch *(Fig. 1-4)*. The digital connections were between the channel banks at both ends of the path. In some instances the channel bank function is built into the switch hardware.

Large numbers of channel banks were installed in central offices, placed between analog switches or between a digital switch and subscribers' analog local loops. Inside the phone company the connection was digital, but that format was unavailable to the user.

POTS on Switched Digital Service

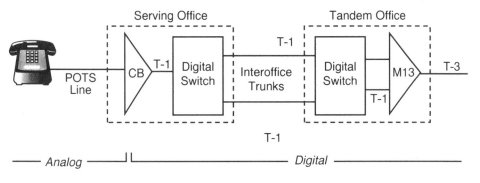

Figure 1-4

ISDN was intended to bring the connection in digital format all the way to the customer site and make digital switched services available in *digital* form.

Digital Local Loops

The worldwide telecommunications industry struggled to define a way to carry a digital signal between customer sites and the switches in the central offices. As is usual in such a search for new technology, several methods were proposed :

- Ping-pong: the local loop is 2-wire, half duplex; both ends take turns making brief transmissions at a speed much higher than the nominal speed of the loop. Time between transmissions was needed to turn the line around, which happened many times per second. This method is similar to an early modem technology.

- Proprietary 4-wire: similar to the 4-wire Channel Service Unit (CSU) technology that is still used today in channel banks to provide DDS local loops at speeds up to 64 kbit/s.

- 2B1Q: the ultimate winner, encodes two bits per baud, that is, every change of state on the line (a 'baud') represents 2 bits. This is possible because the line has four possible quaternary states of ±1 and ±3 volts. Hence the name, 2 Bits (per) 1 Quaternary (one of a set of four, there being four possible voltages sent to the line). By using echo cancellation in the transmitter/receiver at each end, 2B1Q operates full duplex over a single pair of copper wires.

The decision was based on many factors, some political. Technical goals included minimizing cross talk between ISDN circuits in the same cable, avoiding interference with analog voice loops, reducing electromagnetic radiation and susceptibility, and maximizing the length of a local loop over which transmission is reliable. In all respects, 2B1Q is acceptable and has been adopted almost universally. In fact, 2B1Q BRI is exactly what is offered as IDSL, a form of digital subscriber line service.

Some other technologies persist as proprietary products. One was adopted by a Regional Bell Operating Company for deployment in local loops.

An appealing aspect of 2B1Q is that it scales up well. That is, it works at speeds below 64,000 bit/s and has been deployed at over 2 Mbit/s, where it is called HDSL. The nature of 2B1Q, having four voltages, lowers the frequency range that carries the information compared to T-1 coding. Lower frequencies reduce losses which extends the length of usable local loops. Even older technology (1995) allows a single copper pair to carry half or three-quarters of a megabit per second for the maximum distance of a standard local loop: 18,000 feet. By contrast, the "alternate mark inversion" signal for standard T-1 lines needs a repeater (digital amplifier) every 6000 ft. *(Fig. 1-5).*

Limiting each copper pair to a fraction of a megabit/sec means that no repeaters are needed. The outside plant is entirely passive—just wires. Multiple loops in parallel provide faster services: 2 or 3 loops for T-1, 3 or 4 for E-1. Over shorter loops the speed may be increased above 2 Mbit/s on a single pair. HDSL2, a standard that supports T-1 over a single pair, uses a different technology. (For more information on DSL at speeds up to 50 Mbit/s see the author's book *T-1 Networking*.)

So even a traditional T-1 local loop may have three or four active devices scattered over the landscape. Installing and maintaining repeaters is a significant portion of the cost of providing T-1 service. Repeater installation used to causes most of the delay in provisioning new T-1 service. ISDN technology eliminates cost

2B1Q Avoids Active Devices in Loop

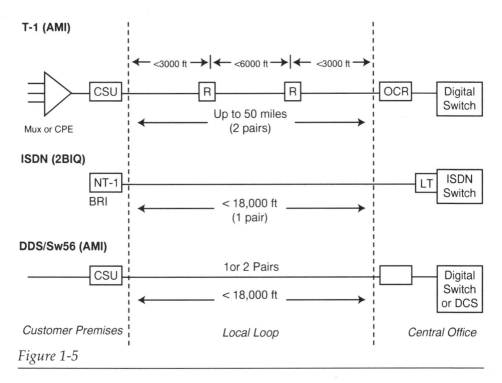

Figure 1-5

and speeds turning up new service. In some countries, 2B1Q is used instead of CSUs on leased DS-0 and DS-1 lines because it is lower in cost. In the US, almost all T-1 lines installed after about 1996 were provisioned with HDSL.

Data Transmission

With a digital local loop, a customer can send 64 kbit/s data in digital format to the network—at least as far as the serving central office (CO). At the switch, problems may arise because of signaling.

Between the first digital CO switches in the US, in the 1970s, T-1 lines carried one voice conversation per DS-0. Most of the time, all the bits represented voice sounds. However, the switches had to tell each other when a phone was hung up or went off hook, what it dialed, and perhaps some additional events.

Robbed Bit vs. Common Channel Signaling

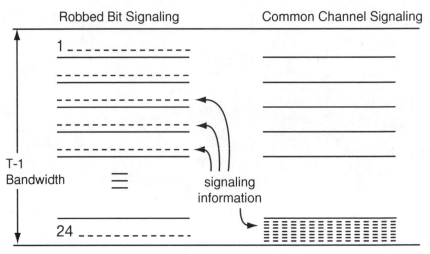

E-1 lines carry signaling in channel 15 (of 0-31).

Figure 1-6

Bell Labs cleverly snuck that signaling information into the same DS-0 that carried the conversation. That way, no matter where or how the conversation was switched through the public network, its signaling went with it. This Channel Associated Signaling (CAS) 'robbed' the least significant bit from every sixth voice sample. Hence the method is also called robbed bit signaling (RBS) *(Fig. 1-6)*.

Changing the least important bit only once in a while had almost no effect on voice quality. However, changing bits in a data stream always has a ruinous effect—the channel's error rate is too high to be useable. To preserve digital data, their bits were not placed in the last bit position, where bit robbing takes place. With data limited to 7 bits out of 8, and with the repetition rate of each sample still 8,000 per second, the data rate became 7 x 8,000 or 56,000 bit/s. When necessary to interwork with older switches, ISDN applies "rate adaption" to mask one bit position, leaving room for 56 kbit/s in a channel. This problem deminished significantly as more central office switches were upgraded to full ISDN capability—most DS-0 connections now run at 64 kbit/s.

European standards put signaling for all voice channels on either an E-1 or 30B+D PRI into a single DS-0 dedicated to signaling. This approach leaves the bearer channels clear for data in all 8 bit positions. Customers in those areas have always enjoyed 64 kbit/s access. The dedicated DS-0 for signaling on an E-1 lines carries signaling bits very similar to the bits robbed on a T-1: but they are multiplexed into one time slot. In ISDN that time slot (numbered 15) performs the same functions, but with a different format based on packetized messages.

Back to Voice (Analog)

So if the entire ISDN transmission system is digital, right into the customer's premises, how can you use a standard analog phone any more? In one sense, you can't—voice must be converted to digital format before it leaves the customer's site. For this function you can buy an ISDN phone. These have always been very expensive because of complexity, small production volumes, and the amount of support needed to install them successfully. Other technologies have made the "pure ISDN" phone an unlikely purchase.

In another sense, you can still use anything you want—there are adapters that digitize the voice and transform the signaling from today's analog phones to the ISDN format. There are "ISDN-ready" PBXs and ISDN terminal adapters for common 2500 telephone sets. So you do not have to replace every phone in the office.

Many Uses for ISDN

ISDN differs from other forms of networking in having many more features available from the carrier service or switch than, for example, a T-1 transmission line or an analog voice circuit. The next chapter gives details of network functions, outlined here.

Plain Old Telephone Service

Lest we forget, the largest use of ISDN for some years to come could well be standard voice communications. When there are ISDN lines on both ends, whether with "ISDN phones" or good quality analog sets and "terminal adapters," the sound is as good as it gets (by today's standard, PCM). Even when an ISDN line talks to an analog line, the quality is often near the top of the scale because almost the entire transmission path is digital. Often, only the local loop at the analog end is not digital.

The International Telecommunications Union—Telecommunications Standardization Sector (ITU-T, formerly CCITT) has defined a higher quality voice encoded with Adaptive Differential PCM (ADPCM) at 64 kbit/s. ADPCM is usually thought of as "compressed voice" intended to run over less bandwidth than a full DS-0, perhaps 32 or 16 kbit/s. But by running at the full 64,000 bit/s ADPCM extends the high frequency limit to over 7000 Hz. This is close to the quality of AM radio broadcasts. It is good enough to handle music at a reasonable quality, though still not "hi fi."

This being ISDN, even POTS is far from plain. In defining a BRI service there are options to indicate how many extensions there are on the line, what kind and how many buttons they have, and whether they are to be considered a hunt group with one phone number or to have individual phone numbers.

Switched TDM channels

ISDN is inherently a switched service, offering circuit-switched or time division multiplexed (TDM) channels on demand. There are other ways to get switched TDM channels, like Switched 56 or Switched T-1, but they traditionally offered only one type of channel per access line: all 56 kbit/s or a full T-1. ISDN allows the user to select at the time of making the connection what bandwidth to connect for that call.

Bearer or B channels are DS-0s. They are available on any ISDN connection; basic rate, primary rate, or faster. The user traffic should 'see' 64,000 bit/s of throughput, but until all switches of all carriers have clear-channel interconnections some calls may be restricted to 56,000 bit/s of usable bandwidth per DS-0. This is done to avoid the bit position where "robbed bit signaling" is inserted.

H channels are larger, available on a primary rate interface or faster connection. They have been defined for several cases *(Fig. 1-7)*.

High Speed 'H' Channels

H_0 384 kbit/s. This rate is also offered by frame relay service providers as an access port speed.

H_1 is the full DS-1, but without the framing overhead. There are options:

     H_{10} 1.472 Mbit/s (North America and Japan), which is the sum of the 23 B channels after devoting a DS-0 to the D channel;

     H_{11} 1.536 Mbit/s (North America and Japan), all 24 DS-0s for which signaling may be sent over a D channel on a different physical interface or in a different DS-1 on the same physical interface;

     H_{12} 1.920 Mbit/s (Europe, Latin America, other areas) 30 DS-0s left after a D channel and one for framing.

H_2 is the full useful payload of a DS-3, which has a line speed of 44.736 Mbit/s (N.A.) or 34.368 Mbit/s (CEPT), including considerable framing and overhead.

     H_{21} 32.768 Mbit/s, 16 E-1s (CEPT) or

     H_{22} 43.008 Mbit/s, the payload of 28 T-1s (N. A.), to 41.160 Mbit/s, if including 18 more DS-0s from the DS-3 overhead

H_3 Would have been 60-70 Mbit/s but so far has been left undefined.

H_4 135.168 Mbit/s (88 T-1s), available on an OC-3 (optical carrier, level 3) access line running at 155.52 Mbit/s.

Figure 1-7

When a network accepts requests on PRI access loops for "H" channel calls, the switches automatically connect multiple contiguous time slots on the calling PRI to the same number of (also contiguous) time slots on the called PRI. The network keeps all DS-0s on the same physical path, which ensures that the timing relationships among them (and the bit order) are preserved end to end. The user's CPE can then treat the group of DS-0s as a single, large time slot.

Inverse Multiplexing

To get a channel faster than 64 kbit/s, and less than a DS-1, the user also may supply CPE that combines multiple B channels via inverse multiplexing. The CPE coordinates the different channels to create what seems to the user to be a single, high-speed channel. Individual DS-0s are dialed up independently. The B channels may be all from the same PRI, or they may be assembled from different PRI or BRI interfaces.

The inverse multiplexer has more work to do than a terminal device attached to an H channel. When the B channels are dialed independently by an inverse mux, each channel is set up as a separate call routed independently through the public network. The physical paths for the channels and their lengths will almost never be the same, resulting in different propagation times. Bit 'A' may be sent before bit 'B' but arrive after bit 'B' because 'A' took a longer path than 'B.'

Inverse multiplexing keeps track of the timing of each B channel, compensating for differences in propagation delay. Thus the bits are delivered in the same order as sent and behave the same as an H channel. Inverse muxes typically transfer data through a V.35 serial interface, but may have direct LAN connections.

Standards exist for three formats for inverse multiplexing:

- BONDING, named after originators, the Bandwidth on Demand Interoperability Group;

- MultiLink Point-to-Point Protocol (MLPPP), defined by the Internet Engineering Task Force in an RFC document; and

- Multilink Frame Relay, an implementation agreement from the Frame Relay Forum.

Due to the drop in costs and increase in demand for bandwidth, fractional T-1 service has moved largely to T-1 access lines. Inverse multiplexing has moved up, to fill the big gap between a T-1 and a T-3.

Interoperability with Switched 56

The earlier digital service called Switched 56 is very similar to an ISDN "bearer service" in some respects. In fact, the same central office switches that provide ISDN access also provide Sw56 service.

There is a 56 kbit/s mode on ISDN that matches the format of a time slot carrying Sw56 on a T-1 line: bit 8 is forced to 1 while user data is being sent. Thus the two services can interoperate across a T-1 interface between digital CO switches. The conversion may also be made within one switch that is providing both services.

Differences lie in the local loop technology. Sw56 uses either a 2-wire or 4-wire local loop. A channel bank with an Office Channel Unit—Data Port (OCU-DP) for Sw56 terminates the end of the loop in the CO. A 56 kbit/s CSU is at the customer site. There is only one channel. Signaling is in-band, and mimics pulse dialing by toggling bit 8 as if going on/off hook (which is what a rotary dial pulser does on an analog phone line).

To set up a connection, one party "dials" the other, in either direction (ISDN to Sw56 or vice versa). When Sw56 is calling, the ISDN switch at the called end tells the customer's network terminating equipment (NT-1, see Chapter 2) that it is a 56 kbit/s call, via a statement in the call setup request message. An ISDN terminal calling a Sw56 user would be told by the network to adopt 56 kbit/s rate adaptation for the call.

Access to Other Services

The switched channel through the ISDN need not go directly to another user's line or terminal. A call can be placed to a node that provides other services or access to them, like X.25 packet switching, Frame Relay Service, cell relay (ATM), or some voice-oriented function like voice mail.

X.25 Bearer Service

The original public data service, X.25 packet switching was standardized first in the 1970s. It stabilized after 1988, the last time CCITT made major revisions in the recommendations. Since then the X.25 interface (which describes how to use the network services, not what goes on inside the network) has become very popular worldwide, though less so in the US.

Users have had two traditional options to access X.25 services *(Fig. 1-8):*

    1. **Leased line** or dedicated local loop from the packet

Access to X.25 Packet Switching Service

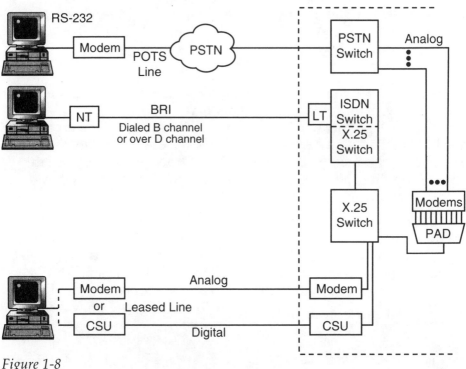

Figure 1-8

switch (in the central office of a carrier, or a privately owned switch) to the access concentrator or Packet Assembler-Disassembler (PAD). Most of these links run at 9600 bit/s, though speeds range from 2400 to 64,000 bit/s.

2. **Analog dial-up**, either from the trunk port of a PAD (via synchronous modems) to a switch port or, more commonly, dial-up from an asynchronous terminal via an async modem to a terminal port on a PAD. The latter is how many on-line computer and Internet services are able to offer local numbers in hundreds of cities: they share an X.25 public network. You may notice that two or more services have the same local phone number in your city.

Most of these connections are upgraded as faster modems become available, and are now (2000) at a nominal 56 kbit/s, though a few locations of some services offer higher throughput for asynchronous access by compressing data in the modems (V.42bis compression).

ISDN offers two more ways to access X.25:

3. **B channel dialup**, usually on a synchronous 64 kbit/s connection. Faster connections could combine multiple B channels via inverse multiplexing, but are not common. The ISDN dial connection replaces a pair of modems and an analog line or service.

4. **D channel connection**, on a virtual circuit using X.25 frames. On a BRI, the throughput available to the user is not very much—about 9600 bit/s. However this is quite enough for short messages. In this case the B channels are not needed at all. Most local exchange carriers offer a "0B+D service" tailored to credit card checking and other point of sale applications. The "always on" nature of the D channel avoids even the short ISDN dial process, making shoppers and shop keepers happy by reducing transaction times to a only few seconds. All the checkout points in an entire store, linked over a LAN, may share the same D channel through router and PAD functions in the ISDN terminal adapter on the BRI line.

There is still a need for the PAD function in the terminal equipment or terminal adapter. The PAD converts data terminal traffic to X.25 format by 'assembling' bytes into packets for transmission on the B channel. Received packets are 'disassembled' to recover and deliver bytes.

When the packet switch is part of the ISDN, it is called a Packet Handling Facility (PHF). It can be part of the main ISDN circuit switch or an adjunct processor that works closely with the circuit switch. Most digital central office voice switches (like AT&T's 5ESS) are packetized (or cell-based) internally and readily handle packets.

Part of the PAD function is to conform to the restrictions accepted when the service was initiated, on throughput, peak data rate, etc. A related job is to negotiate several parameters with the network (actually with the PHF on the switch; see Chapter 3) or with the PAD at the other end of the connection ("triple-X PAD" for async terminal traffic).

Frame Relay Access via ISDN

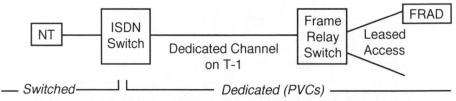

A. ISDN calls to Permanent Virtual Circuits on dedicated FR switch ports

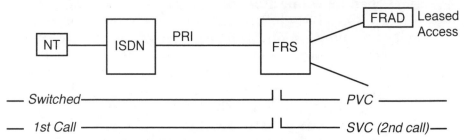

B. FR switch maps Permanent Virtual Circuits (DLCIs) to ISDN call on PRI

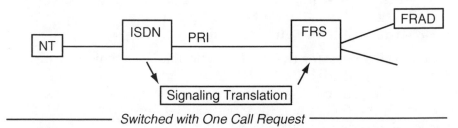

C. ISDN call request sets up access channel and Switched Virtual Circuit

Figure 1-9

Frame Relay

One of the earliest outlines of Frame Relay Bearer Service for the ISDN, CCITT document I.555, described how the circuit-switched ISDN could provide access to a frame-mode service like Frame Relay *(Fig. 1-9)*. The realization of I.555 was delayed until delivery and deployment of frame relay switches that accept ISDN calls and know how to make appropriate frame relay connections. There are several complicating issues related to phone numbers, switch ports, and how customers might share resources in the ISDN and the Frame Relay networks.

Dedicated Access

The simplest way to provide ISDN access to a frame relay service is to dedicate a DS-0 connection, on a T-1 between the ISDN switch and the FR switch, for each user. Then the user, for example on a basic rate line at a remote site, would dial a specified directory number on the ISDN. The call would reach the dedicated port and time slot on the ISDN switch, and thus on the interconnected frame relay switch. One or more permanent virtual circuits configured from that time slot/port into the Frame Relay network could be used by that caller. No one else should call that number on the ISDN because it would connect only to the Frame Relay PVC(s) dedicated to that user.

A slightly more flexible way is to keep the PVCs on fixed time slots, as above. But when the ISDN switch asks for the next available idle time slot, the frame relay switch does not necessarily accept that selection. Instead, the FR switch uses the calling party number in the setup message to look up the user's DLCI and the B channel where it is configured. The FR switch then responds to the ISDN switch with the time slot to use for that particular call.

A public network provider cannot afford to give every user a dedicated channel or port on a FR switch. A large user might need hundreds of ports. FR switch vendors have developed ISDN ports that couple into ISDN signaling (D channels) and make interconnection more efficient, but few users have adopted this technology.

ISDN Call Mapped to PVC

Part of the ISDN incoming call message sent from the ISDN switch asks for a particular B channel. The switch knows which time slots are available, and requests one of them. Delivered with the time slot request is the calling party's directory number. This Automatic Number Indication (ANI) feature is used by call centers (telemarketers) to pull up information about the caller ("screen pop") for the clerk who answers the phone. The FR

---

switch does something similar, using ANI to identify (from a routing table) the permanent virtual circuits (PVCs) in the Frame Relay network that belong to that caller.

The caller's PVCs are then connected within that FR switch to the time slot requested by the central office ISDN switch. A single ISDN call can support many Frame Relay virtual circuits, but in this scenario all PVCs must be activated at once when the ISDN call connects.

ISDN Call Mapped to SVC

Switched virtual circuits (SVCs) are available in Frame Relay networks. Accessing SVC service from the ISDN can work in two ways:

- Two-step calls. An ISDN call, requested via DSS-1 or D-channel signaling, is set up from a frame relay access device (FRAD) to a Frame Relay switch (or PHF). The normal ISDN connection is transparent and terminates there, offering a DS-0 path from the FRAD to the FR switch. The FRAD then requests an SVC from the FR network via an in-band FR message, sent on the control channel (a PVC reserved for signaling). The FR call request results in an SVC through the FR network. The destination may have to be a leased line access device in this method, as FR switches cannot in general dial out.

- Single-step calls. Signaling interconnection between the ISDN and the FR network allows a single ISDN call request from a FRAD to set up a complete connection. Part of the message that requests a bearer channel is passed out-of-band to the frame relay switch. This instructs the FR switch to set up an SVC at the same time that the ISDN network is setting up the switched DS-0 channel.

Standards work needed to define exactly how single-step calling will work appears unlikely at this time. Two-step calling is easily automated and can be implemented now.

Frame Relay Network Parameters

An ISDN connection between a customer's FRAD and a frame handler (frame relay switch) is a simple pipe, a transparent TDM channel. It cannot be configured for committed information rate (CIR), burst rates, or any other FR parameter. The only effect of the ISDN portion of the connection is to limit the throughput to the bandwidth of the channel, the same effect imposed by a leased line.

The frame handler or switch, however, may impose traffic limits. The subscription agreement between customer and carrier will define the conditions. In a private network, the net administrator can allocate capacity by invoking whatever features are supported in the FR switch: CIR, maximum burst size, prioritization, etc.

The rush to the Internet Protocol (IP) as a universal solution eclipsed for some time any development of interworking between ISDN and Frame Relay. However the continued growth in both ISDN and FR use may mean the two could still at some point grow into something closer to the tightly integrated "packet mode bearer service" of X.25 as part of the ISDN.

ISDN Dial Backup for Frame Relay

An interim step toward integration of frame relay and ISDN is to back up leased line access circuits to frame relay service with dial-up connections via ISDN. Products introduced in 1999 made it practical for the first time.

The problem has always existed for large networks, even TDM networks: How do you provide an alternate connection when the remote site loses its leased access? The answer has never been simple.

In a TDM network, it was possible to pair special CSUs, at local and remote sites, that combined leased-service and switched-service ports. For example, two devices, each with a leased line 56 K CSU and a BRI interface, normally operate over the leased line. If that circuit is broken, one box dials the other over ISDN and restores the link *(Fig. 1-10A).*

Consider, however, the network with 5,000 or 20,000 remote sites:

ISDN for Dial Backup

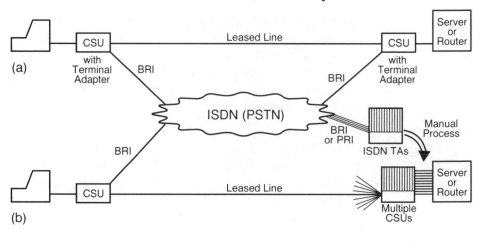

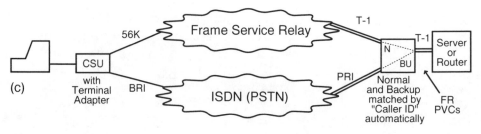

Figure 1-10

this system of dual-port CSUs can't work because it requires 5,000 or 20,000 boxes at the data center. In practice, even a few dozen such connections are unthinkable.

An improvement is to pool a smaller number of CSUs at the data center to be shared by all the remote sites *(Fig. 1-10B)*. Now the problem shifts to how the circuit passing through the node in the data center can be switched from the leased line path to the dial backup path. The TDM node has to find the proper DS-0 (or other small bandwidth circuit) from its host computer and move it to the proper DS-0 on the ISDN port. This feature set never shipped. Changes were done manually, but often took a half hour or more after the remote site phoned in.

Frame relay has a similar problem, to redirect a PVC from the leased to the switched path. The tools needed to solve the problem have existed since at least 1990—they are the access techniques described immediately above. It took a focus on backup to turn out a product, which is what Adtran did with the Atlas 800 Plus *(Fig. 1-10C)*.

A central-site device carries a large number of PVCs from the data center (mainframe, FEP, router, etc.) to the frame relay network. Traffic in and out of the backup unit typically is on V.35 interfaces at T-1. In addition, there are multiple ISDN primary rate interfaces for the backup calls.

Configuration is necessary to match the calling number of each remote site to the DLCI of the PVC going back into the data center. If the PVC between the ATLAS and the remote site is lost for any reason (network problem, access line down, etc.) the remote CSU places an ISDN call to the central site backup unit. The caller ID allows a match between caller and PVC/DLCI, and the logical connection is switched from the normal frame relay access link to the proper DS-0 on a PRI. The failover takes only a few seconds.

Routers at remote sites can perform similar functions by dialing out on separate serial ports through standalone ISDN terminal adapters. However, two drawbacks exist:

1. the routing protocol (RIP, OSPF, etc.) needs time to discover the new physical path, often many minutes, and then needs to announce it to other routers, generating overhead traffic.

2. router ports are more expensive than dedicated backup equipment.

ATM or Cell Relay

ATM vendors are not concerned with ISDN access. ATM over T-1 is defined for a channel speed of 24 times 64,000 bit/s. An ISDN PRI offers only 23 DS-0s for the cell traffic. If ATM access via dedicated T-1 becomes popular, which has happened to a small degree, it is possible that an alternative for PRI will also be defined.

That interface could be as simple as a cell stream at the rate of

1.472 Mbit/s (23 x 64 kbit/s). That rate would require inverse multiplexing, so a more likely rate is an H_{11} (1.536 Mbit/s or 24 DS-0s) via signaling on another ISDN access line. Today there is a definition for up to 20 PRIs to share one D channel, leaving 19 "extra" DS-0s in the trunk group.

With signaling so defined, the H_{11} channel fits exactly the capacity and other specifications for ATM over T-1.

Voice Processing, Etc.

Service bureaus, inside and outside telephone companies, can offer a public voice mail service that is reached by ISDN. More sophisticated systems that record voice messages for later delivery have been working for years. Carriers love voice processing because it generates revenue by completing additional calls and bringing in new per-call fees.

The signaling capabilities of ISDN make many more such services possible. The extent of network functionality will continue to expand as users ask for and carriers deliver additional features.

Signaling Capability

Channel Associated Signaling (CAS) was used from the very beginning of telephony. You asked the live operator for your connection through the same channel that carried the 'connected' voice traffic. Dial pulses really are on the voice line, just an opening and closing of the electrical path. Pulses can operate mechanical relays in the central office to switch a calling line to the called line.

When automated CO switches were introduced, they communicated with each other over analog trunks, in band, with the same pulses. To extend the distance or line length possible between switches, it was necessary to change from current pulses to audible tone pulses. Tones could be amplified easier than pulses (which had to be recreated).

For example, a single high-pitched tone indicates the line is idle—the tone is removed when in use (you may recall hearing a small interval of this tone as a 'chirp' at the beginning or end of a call).

Combinations of two tones were standardized to indicate specific functions, like dialed numbers, which lead to the familiar dual tone multifrequency (DTMF) or push button dialing.

Fairly soon after tone signaling was installed in the public network, "phone phreaks" figured out how to abuse the system. Since local switches don't use the single tone to indicate idle on the subscriber's line, an idle tone coming from the customer was passed to the next switch in the network. If the two switches used tone signaling, the far switch could be fooled into considering the line as idle. When the tone was removed, that switch would accept dialing instructions for a new call. The caller would be paying for the initial call, or perhaps for none at all.

Another recognized problem with in-band tone signaling was that a call request had to proceed from switch to switch as if it were going to be completed. Each stage of call processing used up a voice transmission channel between switches, occupied a path through each switch, and so on. If the called party were busy, or didn't answer, all those connections had to be taken down—without charging the caller. Use of voice paths for signaling got expensive, in terms of needing to build more network capacity than was justified by completed calls.

To avoid fraud and reduce network costs, phone engineers devised a way to send signaling information outside the voice path. The solution was a completely separate, digital, packet switched data network for controlling calls. This was Common Channel Interoffice Signaling (CCIS), later called Common Channel Signaling System (CCSS). There were many versions designed and deployed over the 1970s and 1980s based on X.25 switches linked by 9600 bit/s lines.

CCSS prevents callers from affecting switches by injecting tones because switches no longer respond to tones in the voice path. That's where tones stay; they cannot reach the signaling path.

By coding signaling information in simple messages, CCSS greatly reduced the time needed to transfer a call request. So a request, sent ahead of the voice connection, travels much faster cross-country. Speed kept the CCSS system relatively small for the volume of traffic it controlled, and provided callers with faster connections.

Standardizing CCSS, now known internationally as Signaling System 7, spread the benefits around the world. A call to another continent can be made as quickly as a call within the same state.

The D Channel signaling formats for ISDN access (DSS-1) are similar to Signaling System 7 messages used within the telephone network. This similarity was designed in to allow customers to control some of the features built into the network switches. Adding customer features to the ISDN produces what has become known as the Advanced Intelligent Network (AIN).

Signaling features of the ISDN allow for a variety in services, selectable each time a call is set up. For example, phones capable of higher quality voice or audio can negotiate the use of ADPCM as a bearer channel is connected between them. An incoming call request signaling message can contain not only the identity of the calling party but also the individual extension to ring behind a PBX, the type of information to be sent, and much more.

Where Is ISDN Going?

When ISDN was originally defined in the late 1970s, there was still a hope of circuit switching being "all things to all people" over a single copper pair. There were relatively few communications applications, and practically any of them could work within one or two 64,000 bit/s channels. That speed is much faster than any modem of the 1970s or 1980s, and as fast as or faster than the top speed of digital leased lines of that time (56 or 64 kbit/s). So it was conceivable that a single switched service could replace all analog and digital services.

Unfortunately for the grand plan for ISDN, applications and end users have increased requirements for bandwidth well beyond a DS-0 or even a DS-1. ISDN had planned to grow up as well, into Broadband ISDN based on ATM cell relay service at speeds up to hundreds of millions of bits per second. Like frame relay, ATM pulled away from ISDN, into an entirely separate service. (ATM and broadband ISDN, outside the scope of this book, are covered in the author's *ATM User's Guide*.)

Narrowband ISDN has been expanded to include the full DS-1

or T-1/E-1 line speed. That is, large fractions or an entire 1.5 or 2 Mbit/s link on a Primary Rate Interface (PRI) can be switched using the same signaling format that controls a single DS-0.

The high cost of access loops for 45, 52, and 155 Mbit/s will limit most user sites to a T-1 or E-1 ISDN service for many years to come. Increasing costs of leased lines will make ISDN an appealing alternative. Some local exchange carriers have tariffed ISDN to be less expensive than leased lines when connected only during business hours. For 24x7 connections, leased lines are usually cheaper.

Many vendors of customer premises equipment, like voice PBXs and frame relay access devices, are responding to the growing popularity of ISDN services. The BRI is appearing on new hardware designs, either as the main network port or as a dial backup feature.

Communications features inevitably move into the products they are used with. Modems, and now LAN ports, were once accessories on personal computers. But they have become part of the motherboard. Just as CSUs migrated from stand-alone devices to integral components inside multiplexers, the ISDN capability will migrate from separate network termination boxes to become part of the main product. BRI and PRI will be one of the network trunk options on the list that could include a CSU/DSU, serial port, LAN connection, wireless, infrared, HDSL, or something entirely new. Integration of ISDN is the rule for European PBXs.

The combination of Internet router and ISDN BRI terminal adapter is very popular—use one or both B channels for downloading, take over one for voice on demand, dial into each of two analog voice ports separately (assign one for direct access to a fax machine).

This book explores Narrowband ISDN and the applications that make it a popular technology for access to carrier services.

2.

What is ISDN

ISDN is a digital transmission and switching capability of a network. Standards documents assume it is a public network, but switches with the same functions could be installed in private networks.

The essence of ISDN is the ability to link your network access line to anyone else's access line, and to make that connection on demand. "Anyone else" could be:

- another subscriber to the ISDN,

- a node provided by the carrier for a service like packet switching, frame relaying, or cell relaying; or

- a third party that provides some service like voice processing or credit card verification.

A key component of ISDN BRI is the local loop transmission technology, 2B1Q. This stands for "2 Bits per 1 Quaternary," a way of coding two digital bits into each voltage change on the line (baud). The goal achieved with 2B1Q is to eliminate amplifiers and repeaters in the local loop for digital services. With only copper pairs in the outside plant, ISDN service is easier to install and maintain compared to other services (like traditional T-1) that needed loop repeaters. And 2B1Q makes ISDN available to almost all telco customers on the wires now in place. The result is seen in the low tariffs for BRI access.

While most links between access lines are circuit-switched, a secondary ISDN characteristic (used more outside the U.S.) is to switch a user's packet data on the D channel to some other location on the packet network (which may or may not be connected via ISDN). For packet data service on the D channel, the carrier must have sufficient capacity in the packet handlers that terminate those D channels in central offices. Because that capacity usually isn't in place, D channel data in the US most often handles only short transactions like credit card verification.

Access Interfaces

Two ISDN services have been widely adopted:

- Basic Rate, which carries up to two "bearer" channels for voice or data and a small signaling data (D) channel (2B+D); and

- Primary Rate, based on a T-1 or E-1 interface that carries 23 or 30 voice channels. A full DS-0 is devoted to signaling (23B+D or 30B+D).

Unfortunately the two names (basic and primary) have almost the same common meaning. Think of Basic as being closer to a B channel, so it is the smaller (2B+D). Primary is the other one, the larger capacity access Pipe.

Channels

ISDN uses time division multiplexing (TDM) to create multiple channels on each access link between a user's site and the ISDN switch, usually in a telco central office:

- A "Data Channel" or "D channel" is always present to carry signaling information (call requests, etc.), in the form of packetized messages. The D channel may also carry packets of user data. D channels never carry regular (PCM encoded) voice traffic or circuit-switched user information. A D channel on one interface may be used to control another interface that does not have a D channel.

- "Bearer" channels (of 64 kbit/s as described in Chapter 1) may be provided to carry user information. Usually there are two bearer channels in a basic rate interface (BRI), though one or none is offered by some carriers. ISDN 0B+D is being sold for point of sale devices that use the D channel to verify credit cards, etc. There are 23 (U.S.) or 30 (European) B channels in a primary rate interface (PRI).

- Another channel is always needed for control and synchronization of the local loop. This channel occupies 16 kbit/s in a BRI, 8 kbit/s in a T-1 PRI (the framing bits), and a DS-0 in an E-1 (the same DS-0 used to synchronize the standard E-1 service).

This third type of channel is not available to end users directly, but may be affected by user actions. It is often ignored when discussing bandwidth and access line speeds. You too may choose to ignore it with no operational risk at all.

Signaling

To accomplish its circuit switching, ISDN relies on an "out of band" signaling system, based on data messages sent in the Data or D channel part of the access link. These messages are directed by customer premises equipment to the switch, which may take action on parts of a message and forward other parts to other switches in the network. Except for local calls, it takes multiple switches to complete a network connection.

ISDN out-of-band signaling between switches is Signaling System 7 (SS7, previously known as Common Channel Signaling System, CCSS). It has been used for more than a decade. ISDN switches communicate with each other over SS7, which is made up of 56 and 64 kbit/s links among packet switches (signal transfer points, STPs) devoted to signaling. Packet handlers in ISDN switches may also direct some user data from the D channel on an access loop to a separate packet data network.

ISDN signaling between the customer and the switch is similar, but different: Digital Subscriber Signaling system 1 (DSS1).

US phone lines before ISDN used "in-band" or "channel associated" signaling. That is, the signaling is sent over the same channel as the voice. On analog lines, the channel is the wire pair and it carries either dial pulses, for rotary dialing, or "TouchTone" (technically, dual tone multifrequency or DTMF) for "push button" dialing. When a phone goes "off-hook" (to place a call) the switch returns dial tone. Any pulses or DTMF tones detected by the switch during and immediately after dial tone are considered dialing instructions.

Network Interface Demark Points

The CCITT (predecessor to the ITU-TS) created a reference model for ISDN access loops *(Fig. 2-1)*. The model defines certain points between the customers' equipment and the carrier's ISDN switch. The same model fits both the basic rate and primary rate interfaces.

What happens across each of these Demarcation points is the subject of extensive technical specifications. Even an incomplete library of ISDN-related specifications occupies more than 5 ft of bookshelf. In fact, it is these specifications for functions or the "functional groups" between adjacent demarcation points that define the ISDN.

Functions at the customer premises are assigned to network termination (NT) equipment and/or terminal equipment (TE). The internal structures of the NT and TE are not specified—only the functions they perform and the interfaces to other equipment or the network. Hardware vendors are free to implement NT any way they (or their customers) want. Functional groups may be combined into a single device which might, for example, hide the S/T demark points from the customer.

Each demark point has a specific purpose:

R: Legacy Equipment

The huge amount of pre-ISDN equipment now in operation cannot be changed out immediately. There will be a need to accom-

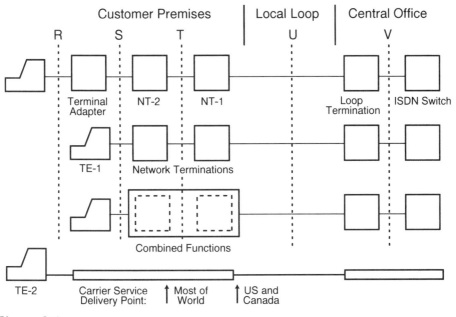

Figure 2-1

modate legacy interfaces for a long time, perhaps indefinitely.

The ISDN model recognizes the need to work with older terminal equipment (TE-2 in ISDN terminology) by offering the R demark. This can be any data or voice interface: RS-232, analog voice FXS, V.35, fax machine, etc. In a sense, these are not part of the ISDN—older CCITT, ISO, and ANSI standards cover them.

The R interface is provided by a terminal adapter (TA) that connects on its network side to an S interface of an NT-2. Some TAs may include the NT-2 functions, which allows them to connect to a T interface on an NT-1. Some TAs will include the NT-1 functions as well, presenting a U interface directly to the network.

How TE-2 output is converted to ISDN formats is up to each vendor. But the resulting format sent to the ISDN network is very strictly defined at the other demark points (S, T, U).

S/T: Customer Premises

When terminal equipment is called "ISDN," or TE1, there is usually an S or T interface on it (some have a U interface). The ISDN phone, PBX, or other customer premises equipment (CPE) may still need an external NT-1 or NT-2 device, or both, between it and the local loop from the carrier.

Outside of the U.S. and Canada, the carrier provides the NT-1, or more. It is the S or T interface (defined later in this chapter) that marks the end of the carrier facilities and the start of a customer's own equipment. It is here that the S/T interface on CPE makes perfect sense—it lets the CPE plug directly into what the carrier provides.

For example, Germany provides an S interface as the standard service. In other countries, it is the T interface that is the demark point. Because of the S interface point, in Europe the Basic Rate Interface (BRI) is known as S0; the Primary Rate Interface (PRI, 30B+D), as S2.

The ISDN specifications as late as 1988 assumed that all customers would have service provided at the S or T interface, not the U interface. In the U.S., by contrast, the Federal Communications Commission prohibits the local exchange carrier (LEC) by from providing customer premises equipment—not even NT-1—as part of the service. When the customer has to provide NT-1, the interface to the network (demark) becomes the U point.

U: Local Loop

The U interface in the US is the demarcation between the public network and CPE. It is here that the physical layer transmission format is standardized on 2B1Q for the BRI.

2B1Q encoding reduces by half the number of "voltage changes per second" (baud) needed to transmit a given number of bits per second (bit rate). That is, the bit rate is twice the baud rate. Baud rate determines signal attenuation, so lowering the baud rate allows the signal to travel farther on local loop copper wires. The beauty of 2B1Q is that it allows the great majority of existing local loops (up to 18,000 feet long, or 1400 ohm loop resistance) to carry more than twice as much traffic in digital form (BRI) as they carry in analog form. This covers 99% of US local loops.

The PRI in North America is based on the T-1 extended superframe (ESF) format with B8ZS (bipolar with 8-zero substitution). The bit rate and baud rate on the PRI are the same (1.544 M/s); each pulse represents one bit. In this same form at the S, T, and U points, two pairs of wire carry 24 channels in the traditional TDM format.

It is ironic that the T-1 signal is more and more often delivered via 2B1Q technology in the form of High-speed Digital Subscriber Loop (HDSL) transceivers, rather than CSUs. As a local loop technology, the U interface with 2B1Q signal format of HDSL offers many advantages for "high speed" data (up to about 750 kbit/s full duplex on a single twisted pair)—even without a real ISDN switched network. HDSL-2 offers the same speeds on a single pair.

Some PTTs use BRI equipment for leased lines of 64 kbit/s. A few US carriers are using the technology to provision higher speed services (T-1 and fractional T-1) without incurring the costs of repeaters in the outside plant. Over short loops, the U interface can carry enough information for compressed video. Some experiments in "dial-a-movie" used ISDN technology.

V: Central Office Connections

There must be some device at the central office to terminate the local loop, generically called the Loop Termination (LT). An LT could be built into the switch. In many digital services (including 56K DDS and T-1) the LT is a specialized device that converts the signal from the loop to a voltage level, pulse shape, and impedance appropriate for distribution within the CO. A T-1 line signal is converted to the DSX-1 format, for example, by an office channel repeater or similar device. In effect, the LT is a CSU.

If the LT for a BRI or PRI is not built into the ISDN switch, then the interface between LT and switch is called the V interface. However, since this point is by definition within a CO and never available to a customer, its detailed specification is outside the scope of this book. Note that this demark point V is not one of the "V series" interfaces defined by CCITT/ITU—those are serial data ports and modems.

Basic Rate Interface (2B+D)

Residential and small business customers find the capacity of the basic rate interface suited to their needs. Two bearer channels provide voice circuits, simultaneous voice and data, or may be combined into a single data channel of 112 or 128 kbit/s. It is common practice to install multiple BRI lines into a location.

Any ISDN service allows for an unrestricted digital connection (any bit pattern, including all 1s and all 0s) at 64 kbit/s, at least locally. Now that interconnections between carriers are almost all "clear channel," only rarely is a call between networks in the US restricted for various reasons, like a transmission line with a 1s density requirement, or the potential for a switch to insert robbed bit signaling into the data stream. These cases limit a user to 56 kbit/s per B channel, rate adapted to the full 64 kbit/s.

In addition to bandwidth for the bearer channels and the D channel, there is also loop overhead for synchronization and testing (maintenance or M channel). Each of the various functions is assigned a channel based on time division multiplexing. Like T-1 multiplexing in the traditional digital hierarchy, the bit flow is organized into frames and superframes. For the BRI, frame format varies by direction of transmission and demark point.

U Interface of the BRI

In most of the world the ISDN customer never sees the U interface. The carrier provides the NT-1 device so the service is based on the S or T interface.

In the U.S. and Canada, the carrier presents the U interface on an 8-position modular jack on the end of the local loop *(Fig. 2-2)*. Customers furnish an NT-1 that meets the U-interface specifications (also at a modular jack). An 8wire (4-pair) cord with two modular plugs connects them. NT-1 may be powered locally on the U connector.

Basic Rate Access (BRA) was designed to operate over existing analog voice-quality local loops. There is supposed to be no need for special pair selection, conditioning (removal of bridged taps), etc., for loops up to about 18,000 ft. long. Loading coils used to shape frequency response on long loops must be removed. The

'U' Interface Modular Connector Pinout

Pin[1]	Color[2]	Function
1	Green	Battery Status Indication
2	Green/White	Battery Status Indication
3	Orange/White	No connection (reserved)
4	Blue/White	Local Loop, either conductor (polarity not observed)
5	Blue	Local Loop, either conductor (polarity not observed)
6	Orange	No connection (reserved)
7	Brown	− Recommended Power Supply (PS-2, 48 V DC nominal)[3]
8	Brown/White	+ Power (this lead grounded if power not floating)

[1] Pins numbered left to right when looking into the jack cavity with locking clip down.

[2] Dominant color/stripe color; twisted pairs may have complementary colors (B/W, W/B) or solid color on one strand and same color with white stripe on the other.

[3] Power transfer may have many configurations; see Chapter 5.

This wiring based on TIA-568A, recommended for ISDN. TIA-568B (same as AT&T 258A) swaps pair 2 (pins 3 and 6) with pair 3 (pins 1 and 2), changing only the color of the wires on the pins. Electrical performance is the same. Different connector vendors may indicate different colors.

Figure 2-2

design limit is a signal loss of about 42 dB from the original signal of 13 to 14 dBm.

The ability to send more than two digital channels in place of one analog conversation depends on the way the digital signal is encoded.

2B1Q Line Coding

The BRI line signal from the network is coded in "2B1Q" *(Fig. 2-3)*. Each pair of consecutive bits is coded into one of four (Quaternary) values. "Quat" is shorthand for a voltage level and the two bits it represents. There are four different quats to denote the four possible pairs of bits, numbered for reference as ±3 and ±1. Their nominal voltages are defined in a 3:1 ratio, specifically at ±2.5 and ±5/6 V at the output of a transmitter.

2B1Q Line Coding

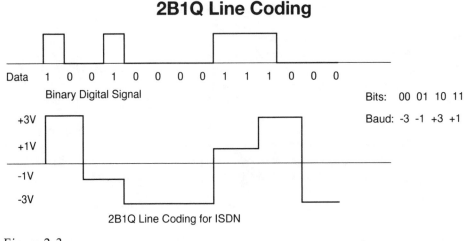

Data   1   0   0   1   0   0   0   0   1   1   1   0   0   0

Binary Digital Signal

Bits: 00 01 10 11

+3V

+1V

Baud: -3 -1 +3 +1

-1V

-3V

2B1Q Line Coding for ISDN

Figure 2-3

This format is also called pulse amplitude modulation (PAM) because the size of the pulse conveys as much information as its polarity. The two bits in each quat, from this viewpoint, are the sign (polarity) and magnitude of the transmitted pulse.

An oscilloscope trace won't show the translation of bits to quats exactly as outlined for two reasons:

- the bits (except for synchronization words, see below) are scrambled ("with a 23rd order polynomial," which sounds like a lot) to break up long runs of 1s or 0s. The receiver deframes, decodes quats to bit pairs, then descrambles to get the original data.

- loop current (1 to 20 mA) may flow on the same wire pair, perhaps to power NT-1. Almost always a sealing current is used to prevent corrosion of electrical contacts in the local loop. Pulses are superimposed on the steady current. The receiver looks for the pulses and ignores the loop and sealing currents.

Both the central office and the customer equipment transmit at the same time, full duplex, over a single wire pair. A "hybrid" circuit, similar to the one in an analog telephone, couples both transmitter and receiver to the same wire pair. Loop current passes through this circuit, and may be fed to the rest of the NT-1 for its power.

Full Duplex Transmission with Echo Cancellation

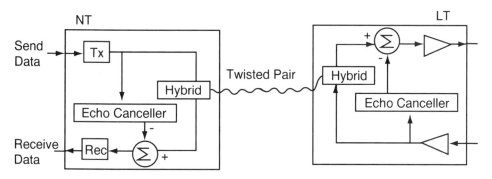

Σ = summation function subtracts local signal from hybrid output.

Figure 2-4

The receiver at each end listens to the signal on the wire, but subtracts from it the signal being sent by its own transmitter *(Fig. 2-4)*. The difference is the signal from the far end.

The subtraction process involves serious digital signal processing for echo cancellation. EC must remove not only the immediate feedback (through the hybrid), but also reflections from points along the local loop caused by changes in wire gauge, splices, bridged taps, or other impedance changes. On startup, the EC at each end learns the properties of the loop it is on and adjusts automatically. This entire function for the BRI has been reduced to a single integrated circuit chip (plus external transformers) by several manufacturers.

The D channel of 16 kbit/s, plus a total of 128 kbit/s for two B channels, provides a bandwidth of 144 kbit/s available to the user (though not all at once). Another 16 kbit/s (in the form of synchronization words and overhead or maintenance channels) brings the gross bit rate to 160 kbit/s. Encoding with 2B1Q results in an 80 kilobaud line signal.

That clock rate will be within ±5 ppm, coming from the network, under all but the worst conditions of network testing or disruption. Normally the network will be driven by a Stratum 1 clock source accurate to 0.0001 ppm or better. After an NT synchronizes to the received data (loop timing), it will transmit to the

Bidirectional 2B+D Frame at BRI 'U' Interface

2B+D	1			2				12		
Quats	1-4	5-8	9	10-13	14-17	18	...	99-103	104-107	108
Bits	1-8	9-16	17-18	19-26	27-34	35-36		197-206	207-214	215-216
Channels	B1	B2	D	B1	B2	D	...	B1	B2	D

Figure 2-5

network at exactly the same rate. When running free, NT should be within ±100 ppm.

Framing

Like any other TDM transmission, the U interface is framed. A frame of 2B1Q encoded user information at the BRI consists of a synchronization word, twelve octets from each bearer channel, 3 octets from the D channel, and 6 bits of overhead. That is, each frame contains 12 sets of 2B+D samples *(Fig. 2-5)*. B1 and B2 are the two bearer channels. Two consecutive D bits are sent (as one quat) after each pair of B bytes from the bearer channels.

Bits received from the S/T interface are sent in the same order at the U interface. When transmitting voice, the most significant bit (MSB) of the octet (PCM encoding) is sent first.

A synchronization word (SW) of nine pulses of the larger (higher voltage) "3" quats (+ + − − − + − + +) is added at the beginning of each U-frame to ensure the receiver can locate the frames and thus the B and D channels. Eight consecutive frames are organized into a superframe *(Fig. 2-6)* before they are sent to the network. The start of the superframe is marked by one inverted sync word (ISW) to locate overhead fields. In an ISW, each pulse is the opposite polarity compared to the SW, or − − + + + − + − −.

Despite the care taken to locate the B channels on the local loop, they are handled independently from the switch onward into the network. Even if both B channels from a BRI are switched to the same remote BRI, the phase or timing between the channels generally is not preserved. Aggregating two B channels to create a single channel of 112 or 128 kbit/s requires additional processing in the CPE at both ends.

Superframes at BRI 'U' Interface

Superframe transmitted from Network to NT-1

	Frame	2B+D	Overhead / Maintenance					
Quats	1-9	10-117	118			119	120	
Bits	1-8	19-234	235	236	237	238	239	240
Frame #	Sync	Data	M1	M2	M3	M4	M5	M6
1	ISW		Address			act	1	1
2	SW	2B+D per Frame	Data/Msg			dea	1	Febe
3	SW		Embedded Operations Channel			1	1 msb	
4	SW				8	1		
5	SW		Address			1	CRC	
6	SW		Data/Msg			1		
7	SW		Embedded Operations Channel			uoa		
8	SW				8	aib		lsb 12

Superframe transmitted from NT-1 to Network

	Frame	2B+D	Overhead / Maintenance					
Quats	1-9	10-117	118			119	120	
Bits	1-8	19-234	235	236	237	238	239	240
Frame #	Sync	Data	M1	M2	M3	M4	M5	M6
1	ISW		Address			act	1	1
2	SW	2B+D per Frame	Data/Msg			ps1	1	Febe
3	SW		Embedded Operations Channel			ps2	1 msb	
4	SW				8	ntm		
5	SW		Address			cso	CRC	
6	SW		Data/Msg			1		
7	SW		Embedded Operations Channel			sai		
8	SW				8	R(=1)		lsb 12

Shading marks differences between transmission directions.
Frame duration is 1.5 ms.

Figure 2-6

Control/Overhead

The last six bits of each frame (M bits) carry the overhead and control fields of the superframe. This bandwidth is sometimes called the M channel (as in 2B+D+M) for Maintenance. M bits have no meaning in a frame alone; they rely on the superframe for definition. Some fields (EOC and CRC) overlap multiple frames, while the function of the M4 position changes from frame to frame. Some conditions at U are mapped by the NT-1 to other control signals at the S/T interface (see next section).

Note that the content of the control fields differs for each direction of transmission.

CRC A 12-bit cyclical redundancy check (CRC12) is calculated from the B, D, and M4 bits of a superframe. The value (check sum) is inserted into the next superframe. The receiver makes the same calculation (usually in a hardware shift register) on the same fields in each superframe, and saves the value to compare with the CRC received in the next superframe.

FEBE If a receiver finds the CRC in the superframe differs from the CRC calculated from the previous superframe, an error is declared. To inform the sender of the error, the receiver changes the far end bit error (FEBE) bit in the next sent superframe from 1 to 0.

ACT In both directions, the ACTivate control bit is used for start up. When it is ready to communicate at layer 2 (link layer) the network sets ACT = 1. NT-1 uses ACT to alert the network when there is traffic to send after a standby period.

DEA Where the central office supplies power to the NT-1, the network may deactivate or turn off the NT-1 to reduce power consumption. By resetting DEActivate from 1 to 0 the network tells the NT-1 to go into low-power standby mode. The user can't turn off the network, so there is no DEA field in the superframes sent from the NT to the network.

PS In the U.S., the customer furnishes the NT-1 and the power to run it. Overhead is not directly accessible by the customer, except for the power status bits (PS1 and PS2). As described in Chapter 6, a local power supply with a battery backup may report the battery condition and presence of commercial power to the network here. NT-1 translates that information into the PS bits: 11, normal; 10, backup is out; 01, primary power out, running on backup; 00, a "dying gasp" to alert a network manager that total failure has occurred and operation may cease.

NTM When a customer initiates a test mode in the NT-1, it advises the network by making the NT Test Mode bit (NTM) a 0. During normal operation NTM is 1.

CSO Starting up the link between NT and network for the first time begins from a "full reset" or a cold start and may take as long as 15 seconds. In the process, the two ends "tune" their echo cancellers and other parameters to attain a low error rate. Many of those parameters may be remembered, and used immediately after a standby period or when power is recovered after a loss. Some NTs have to start from "full reset" each time. They set the CSO (cold start only) bit to 1. NTs that can recover more quickly (in less than 300 ms) set CSO to 0.

UOA The ACT bit normally activates both the U interface and the T or S interface behind it. If the network doesn't want to activate S or T, it sets UOA (U interface Only Activation) to 1. This inhibits the S/T interface from starting up.

SAI In the opposite direction, activity at the S/T interface causes the NT to set the SAI bit (S/T Activity Indicator) to 1.

AIB The alarm indication bit is set to 1 to tell the NT that a path exists as far as the first switch in the serving central office. The transmission equipment between the CO and the NT-1 will set AIB = 0 when the path is interrupted. A common reason for interruption is a loopback test in progress.

R The position corresponding to AIB in superframes to the network, and all other positions not specified, are reserved for future use (to be standardized). They are set to 1 before the information is scrambled and encoded in quats.

EOC The embedded operations channels are used by the network to command the NT-1 device for management and testing. The NT-1 only reacts to or echoes the commands from the network in normal operation, but must initiate messages on power-up or when a TE requests service.

The 3 address bits can deal with 7 locations along a transmission path, but the NT-1 responds only to its own address (000) or the broadcast address (111). The Data/Message bit indicates whether the byte in this superframe carries a command (=1) or numerical data (=0).

The commands are 8-bit words, only 8 of which are standardized *(Fig. 2-7)*. Two additional block of 64 codes each are set aside, for non-standard uses and for internal network functions. All other message values are reserved (not to be used until standardized). In normal operation, the NT echoes received mes-

Operations Messages in the EOC

Direction NT Ntwk	Message Code	Function
<——	0101 0000	Loopback of both B channels and D channel
<——	0101 0001	Loopback of first B channel alone, transparently
<——	0101 0010	Loopback of second B channel alone, transparently
<——	0101 0011	Asks NT-1 to send corrupted CRCs, for testing
<——	0101 0100	Warns that network will be sending corrupted CRCs
<——	1111 1111	Return to normal (unlatch all test modes)
<——	0000 0000	"Hold" state: freezes NT-1 EOC processor
——>	0000 0000	NT received message with address not 000 or 111
——>	1010 1010	Unable to comply: function not supported, etc.

All commands "latch ON" a function, which is released by "Return to Normal." Dedicated blocks of codes are assigned for proprietary uses (codes starting with 0100, 0011, 0010, 0001) and internal network use (0110, 0111, 1000, 1001).

Figure 2-7

sage codes in the next available superframe except when:

- a command is not recognized or not acceptable; the NT echoes no more than two copies of the code byte, then returns the Unable to Comply message.

- a valid message code arrives with an address other than that of the NT; that is, not 000 or 111. NT then returns Hold (all 0s).

An NT does not act on a command until it receives three consecutive superframes with the same message code in the EOC. The network sends the same message code repeatedly until acknowledged by the NT. Acknowledgement is an echo of the same code in three consecutive superframes. The network may then continue to send either the same code or the hold message.

Start Up Procedures

When an NT-1 is first installed on a line (plugged into the U interface), and after a power off/on cycle, it is responsible for starting up the connection to the network. This is a cold start, in which the equalizers and echo cancellers are adjusted to the characteristics of that particular loop. The signals used in startup *(Fig. 2-8)* prevent user data from passing until the connection is

Startup Signals at U Interface of BRI

Signal Name[1]	Synch Word	Super-Frame	2B+D Fields[2]	M-Bit Values[3]	Duration[4]
Unframed Wake Up Tone, Not Scrambled					
TN					6 frames
TL	Alternating "3" quats: ++++– – ++++– –....				2 frames
Startup and Normal Frames and Superframes					
SN1	present	no ISW	1	1	T1 - T2
SN2	present	no ISW	1	1	T5 - T6
SN3	present	present	normal	normal	> T6
SL1	present	no ISW	1	1	T3 - T4
SL2	present	present	0	normal	T4 - T7
SL3	present	present	normal	normal	> T7

[1] L and N indicate source of signal: NT at customer site or LT in central office.
[2] All three channels, before scrambling. Normal is user data.
[3] Normal implies all EOC and control bits are functioning.
[4] Times are identified in next Figure.

Figure 2-8

ready and goes "transparent."

The steps taken in starting up the local loop connection are outlined in *Fig. 2-9.* The signals are defined in *Fig. 2-10.*

What the previous table describes is the generic startup, for both cold and warm starts. Here it is clear why "cold start only" is not a desirable feature in an NT: it could take as long as 15 seconds to adjust the echo canceller and equalizer for the line. If the NT reports that it is Cold Start Only, the network should not turn it off by setting DEA = 0. The 15 s for a cold start is too long to wait before every phone call even starts to go through, so deactivation is not used in that case.

On first application of power to NT-1, it should attempt to do a cold start, then remember the line parameters in order to allow warm starts from then on. Warm starts take less than 300 ms.

NTs return to Reset mode (silent, listening) when they do not receive a signal or lose synchronization for more than 480 ms, or

BRI Startup Procedures

Handshake Procedures

Time Point	Interval Starts/Ends		Signal From NT	Signal From Ntwrk	Status
T0	any period		none	none	Both ends in Reset mode, not transmitting
| any	A		none	TL	Network initiates wake up tone
| <4 ms			TN	none	NT-1 responds with wake up tone
or | any			TN	none	NT-1 initiates wake up tone
T1			[SN1]	none	[Optional] signal
T2	B	A	none	none	NT stops TN or SN1
<480 ms			none	none	NT awaits response from network
T3			none	[SL1/SL2]	Network responds to end of NT signal, SN1
T4	C	B	none	SL2	Network indicates ready to receive, SN2
T5			SN2	SL2	NT has acquired bit and SW frame sync; equal- izer and echo canceller converged
T6	D	C	SN3	SL2	NT has superframe sync; now operational
T7	any	D	SN3	SL3	Network (LT) has superframe sync; B channels transparent in both directions

Timing Constraints

	Cold Start	Warm Start	Time Allowed To:
A+C	<5 s	< 150 ms	NT
B+D	<10 s	< 150 ms	Network

Figure 2-9

Signals to Trigger BRI Maintenance Modes

Number of Pulses	Function Name	Effect
6	NT Quiet Mode	Stops NT-1 from sending for 75 seconds
8	Insertion Loss Measurement	NT sends test frames (e.g., SN1 or SN2)
10	Exit Maintenance Mode	Cancels test
1-5 7 9 >10	none	NT takes no action

Figure 2-10

when ordered to turn off by the network (DEA=0). If a call comes into the NT, the network turns it on with wake-up tone Data from the S/T interface (specifically, receipt of the INFO3 signal, see below) will activate the NT, which will start up the link, also via the wake-up tone.

Maintenance Modes

Even amid the latest digital loop technologies, like 2B1Q, an old reliable signaling technique crops up again: emulated rotary dialing. That is, the central office can put the NT-1 into test modes by opening the local loop to interrupt the 48 V battery current in the circuit. Pulses are 40-60% on versus off, at a rate of 4 to 8 pulses per second. Rotary phones usually dial the same way, but slightly faster at 10 pps. Like a phone dialing a digit as a group of that many pulses, this control signal must be preceded and followed by a half second or more of steady current.

For those installations where there is no sealing current in the loop (current below 200 microAmp), the CO imposes electrical pulses instead of interrupting the battery current. The applied signal is a slow (2 to 3 Hz) 60 V peak sine wave. Each half cycle counts as a pulse.

A timer automatically takes the NT out of test mode after 75 s. Repeating the pulse train before expiration of the timer restarts the timer.

Specialized test equipment in the CO generates these signals. An old phone pulser might do the job in a user's lab, if a human finger slowed the wheel as it rotated back. Making 2 Hz sine waves requires a capable signal generator with sufficient power.

Only the three pulse trains defined in the Table are valid.

Quiet Mode (QM) prevents the NT from transmitting for any reason. QM allows the CO tester to test loop continuity and to exercise the non-linear circuit in the NT-1. Based on a semiconductor, this "metallic termination" draws very little loop current until the voltage at the NT rises above about 35 V. At that point it switches on and draws current. It stays on until loop current falls below 0.1 mA for at least 100 ms. This behavior is characteristic of an NT-1 and can be used by the phone company to check for a standard NT-1 on a specific loop.

An **Insertion Loss Measurement Test** (ILMT) requires the NT-1 to transmit valid looking superframes (with scrambled, random data) at an average power level of 13.5 dBm. Measuring the power received at the CO shows the loss over the loop.

Exit from Maintenance results from the receipt of 10 pulses. When the timer expires, or the NT receives 10 pulses, it attempts to restart the connection with the central office in a normal cold or warm restart.

S/T Interface of the BRI

Keep in mind that NT-1 and NT-2 are "functional groups" that need not be separate physical devices. At each interface a functional group will provide bit timing (clocking recovery, based on the signal received from the network); framing (from unique bit patterns); delineation of B and D channels and octet timing (based on framing); D channel access procedures (signaling); and power feeding.

Wiring Topology

Much of the difference between S and T is in the number of devices that may be attached to each.

- T is a point-to-point connection, between NT1 and NT-2, consisting of two balanced 'interchange circuits' of a single

Multiple TEs at the S Interface

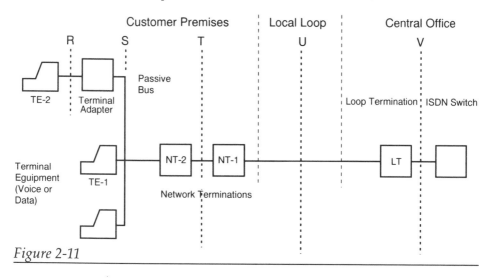

Figure 2-11

copper pair each. Polarity of each pair is not significant. NT-2 may be built into CPE and need not be a separate device.

- S may be Pt-Pt, but also supports a "passive bus." The physical layer for this bus interface *(Fig. 2-11)* is two twisted pairs of copper, one for transmission in each direction. Because it connects multiple terminal devices in parallel, polarity of each pair must be consistent at each connector. Two additional pairs may provide power and power monitoring *(Fig. 2-12)*.

Polarity of the two other wire pairs must be maintained, in case they are needed to deliver d.c. power. Reverse power polarity indicates the source is the backup or reserve supply and may inhibit some TE functions. Local wiring is detailed in Chapter 6.

The transmitter of the NT-2 toward the terminals may have several receivers attached. All of the terminals' transmitters (one per terminal device) connect to the single receiver on the NT-2. There are procedures (see below) at the S point to control access to the B and D channels by multiple terminals.

See Chapter 5 for details on inside wiring.

'S/T' Interface Modular Connector Pinout

Pin[1]	Color[2]			Function
1	Green	]	+	(this lead grounded if power not floating)
2	Green/White	]	−	Power Source 3: sink on NT; source or sink on TE
3	Orange/White		+	side of pair transmitting to network (TE to NT)
4	Blue/White		+	side of pair receiving (NT to TE)
5	Blue		−	side of pair receiving (NT to TE)
6	Orange		−	side of pair transmitting to network (TE to NT)
7	Brown	]	−	Power Source 2 (PS-2, 48 V DC nominal)[3]
8	Brown/White	]	+	(this lead grounded if power not floating)

[1] Pins numbered left to right when looking into the jack cavity with locking clip down.

[2] Dominant color/stripe color; twisted pairs may have complementarycolors (B/W, W/B) or solid color on one strand and same color with white stripe on the other.

[3] Power transfer may have many configurations; see Chapter 5.

This wiring based on TIA-568A, for ISDN. TIA-568B swaps pair 2 with pair 3, changing only the color of the wires on the pins. Electrical performance is the same if all connectors are wired consistently.

Figure 2-12

Information Framing

Again, the TDM structure of the S/T interfaces is defined by framing. The B channels are byte interleaved, which means octets of each bearer channel alternate in the transmission to minimize latency. The 16 kbit/s D channel is interleaved as four bits spread out among the four B channel octets in a frame *(Fig. 2-13)*. Note that the frame structure is different for each direction, to or from the network, though frame lengths are the same.

S/T frames are 48 bits long. From the presence of two B channel positions (16 bits), these frames must have a duration of 250 microseconds (one frame in 1/4000 s) to support 64 kbit/s per channel. Dividing 48 by 0.000250 s gives 192,000 bit/s.

So, not only are the S/T frames different from the ones at the U interface, the bit rate is different as well. One of the functions of the NT-1 is to adapt the two bit rates so the B and D channels

Framing at the ISDN S/T BRI Interface

Frames transmitted from Network Termination to Terminal Equipment

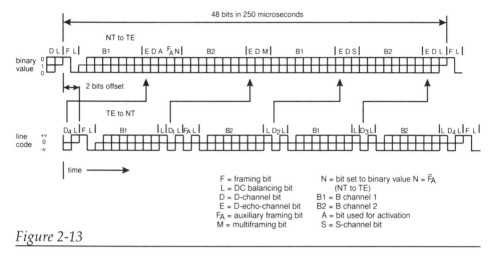

Figure 2-13

pass transparently. It does this by deriving the 192 kbit/s rate from the 160 kbit/s clock extracted from the U interface. When free running, the clock rate should be within 100 ppm.

In sending to the NT, the TE starts its frames nominally 2 bit times after the times of frames received from the network. Transmission or propagation delay can alter this relationship where the frames are received at the NT.

The electrical signal (the same in both directions) resembles "alternate space inversion," to borrow from the language of T-1 networking. The mask of the pulse shape calls for a square wave, but with acceptance of a large spike overshoot on the leading edge and some undershoot allowed on both sides. 2B1Q encoding of bit pairs in quats (standard for the U interface of the BRI) does not appear at the S or T interfaces.

Recall that a T-1 line sends 0s (spaces) as no voltage, 1s (marks) as a voltage whose polarity reverses with each successive "1" sent (alternate mark inversion, AMI). ANSI calls the S/T interface format "pseudoternary coding," but it is almost the same electrical signal as AMI *(Fig. 2-14)*. The differences relate to the framing (F) bit always being positive and the requirement for

---

Pseudoternary Bit Coding at the S/T Interface

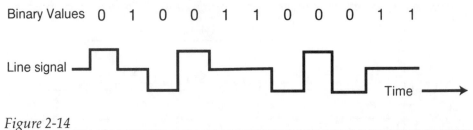

Figure 2-14

certain bipolar violations (two pulses of the same sign, rather than in the alternating pattern).

When a B channel is idle, the TE sends binary 1s (no pulses) to the NT. A quiet TE allows another TE to use the shared bus.

The logical interpretation of the ISDN signal is close to a T-1 with "inverted bits" (1s changed to 0s and vice versa). The specification for the S/T interface, however, requires a pair of "bipolar violations" at the start of each frame. A BPV is two pulses of the same polarity, not following the alternating polarity rule. On point-point S or T connections, the receiver doesn't care if the wires are reversed (F may be sent positive and received as negative).

TE to NT F itself, always positive, creates a BPV because the preceding L bit is a positive pulse if and only if the last data bit is not a positive 0 pulse. Second, the first data bit after an F bit must be negative, the same polarity as the preceding L bit. These two BPVs per frame help the receiver locate and confirm the F bit to establish framing. A receiver that cannot locate framing will declare a loss of frame alarm after three frame times. Three normal sets of BPVs are enough to assume framing is recovered.

NT to TE F produces a BPV, same as in the TE to NT direction. The auxiliary framing bit FA produces the second BPV. Only two frame times are needed to lose and regain framing in this direction.

ANSI also adds an inverted mode that is very close to AMI, but with the additional requirements for the framing and balancing bit patterns.

PCM voice encoding will not produce many successive octets of all 1s, as this is an extreme limit of loudness which must soon change to the opposite polarity. However, to ensure that an all 1s signal is acceptable, the bits are "scrambled," in a controlled way, to improve the statistical probability of an equal number of 1s and 0s on the line.

Frame Overhead for Control

Additional channels occupy another 12 bits in each S/T frame *(Fig. 2-15)*:

- F (framing bit) is always the same polarity (+) and followed (within 13 or 14 bits) with a bipolar coding violation by the first 0 in the B1 bits (to make F easy to find).

- L (Last bit), like an even parity bit, equalizes the number of + and - pulses in the section of frame between L bits. There are multiple L bits, each the last bit in a group, to balance the net average current in a pair of signal wires to zero. For example, the L bits following F and D, to balance those single bits, are always the same binary value (if D=1 then L=1, if D=0 then L=0).

- A (Activation bit) sent to the terminal to wake it up from a low-power mode or to put TE to sleep (useful to carriers who are supplying power to, for example, telephones). While not required, Bellcore recommended that NT and TE power down when not active with a call on regular power, but especially when on standby power (battery backup).

- FA or auxiliary framing bits from TE to NT may be used as a Q bit in every fifth frame (see Multiframe Overhead, following).

- D-channel bits contain the signaling messages. This channel is shared where there are multiple TEs.

- E (echo) bits return the D channel traffic received at the NT from any TE, back to the TE(s).

The procedures used by both NT and TE on the D and E channels control access to the D channel when there are multiple TEs. The method depends on manipulating priorities for each TE device, to let all TEs send a signaling or data class message frame

Summary of S/T Control Bits

Bit Position	Bit Name	TE to NT		NT to TE
1	F	Framing bit (Line Code Violation, LCV)		
2	L	Balance bit		
3-10	B1	First Bearer Channel (first octet); 1st "0" bit is LCV		
11	L	Balance bit	E	Echo of D channel
12	D	D Channel		
13	L	Balance bit	A	Activation
14	F_A	Auxiliary framing (LCV if not Q bit)		
	Q	Every 5th frame, if used		
15	L	Same binary value as F_A	N	Binary value opposite of F_A
16-23	B2	Second Bearer Channel (first octet)		
24	L	Balance bit	E	Echo of D channel
25	D	D Channel		
26	L	Balance bit	M	Multiframing; 1 only in frame 1
27-34	B1	First Bearer Channel (second octet)		
35	L	Balance bit	E	Echo of D channel
36	D	D Channel		
37	L	Balance bit	S	Status, power loss
38-45	B2	Second Bearer Channel (second octet)		
46	L	Balance bit	E	Echo of D channel
47	D	D Channel		
48	D	Balances D bit (47)	L	Balances frame

Line Code Violation; bit is same polarity as preceding bit, not in alternating pattern.
Binary Value refers to 1/0, not polarity (framing does not depend on polarity of pulses).

Figure 2-15

before an individual TE has a chance to send another message of the same class.

Between message frames, the NT and TEs send all 1s on the D channel (both directions). Recall that the pseudoternary line encoding converts 1s to an absence of pulses—the channel is quiet so any number of TEs may be on the line and not interfere with each other. Because the NT is the only transmitter sending downstream (toward TE), it may send HDLC flags (01111110) instead of all 1s.

On the E channel the NT returns to the TEs any signal received from them on the D channel. When there are no messages, E bits are all 1s, or no pulses, the same as the D channel.

TEs monitor the E channel, counting the number of 1s (bit times with no pulse) since the last 0 (pulse). The count starts over every time a pulse is seen. Each TE's priority is set through this count (called C). A TE with a higher priority can start to send when C is lower (sooner) than a TE with a lower priority (which must count to a higher value of C before starting to send).

Signaling frames have a higher priority than user data, and within each class there is a higher and lower priority. This leads to four significant values for the count:

- Signaling, 8 and 9;
- All other frames, 10 and 11.

A given TE at startup will assume it has the higher priority in each class. That is, it may start to send a signaling message after it sees eight 1s on the E channel. However, after it has sent a layer-2 signaling frame, its priority for signaling messages drops to "lower" and it must wait for nine 1s. This gives other TEs a chance to exercise their "higher" priority for signaling messages and transmit when C next reaches 8. The same is true for user data frames, with C at 10 or 11.

Since a TE with a "higher" priority will start to send a queued message when the count reaches 8 or 10, the count will not exceed that value until all queued messages in all TEs have been sent and all TEs are in the "lower" priority. So when the count reaches the value for the "lower" priority (9 or 11), all TEs know that every one of them has had a turn and the priority for that class is returned to "higher" at every TE.

While a TE is sending on the D channel, it continues to monitor the E channel, comparing received bits with those being sent. If they are not the same, the TE assumes another TE has attempted to send a D channel message at the same time. The TE that detects a collision immediately stops sending, reverts to listening to the E channel, and counts 1s again looking for its priority level. Some implementations introduce a random length delay to spread out restart times and reduce the chance of additional collisions.

Multiframe S/T Overhead

Multiframes allow two optional maintenance channels, S and Q, to be defined (as they are by ANSI) between the NT and TE on the S/T interface. If the optional Q bits are not used, then multiframing is not required. National ISDN-1 and -2 recommend that S and Q channels be implemented, and that certain conditions on the U interface be mapped by NT-1 to specific signals on the S/T interface.

Yamaha chip designers produced a BRI chip that handles the physical layer (1) and the data link layer (2) in hardware, including access to the S and Q bits from an external microprocessor. However, one vendor refers to the S channel as "reserved for future use."

The S/T multiframe consists of 20 S/T frames. The M bit marks the start of the multiframe when it is a binary 1; in frames 2 through 20 of the multiframe M is 0.

- **Q bits** occupy the position of the FA bits in frames 1, 6, 11, and 16 in the TE to NT direction only.

  In one interpretation, the FA bit, normally 0 (a negative pulse) is a 1 (no pulse) in those positions of a multiframe. You might also think of the Q bit as a form of robbed bit signaling.

- **S bits** are defined in each frame sent from NT to TE, as five subchannels of four bits each.

Standardized functions for S and Q include power status, loopback commands and requests, "loss of received signal" alarm, and similar maintenance jobs. The "user" or TE can signal on the Q channel to initiate self-tests in the NT and to invoke or release loop backs on the B channels.

Bellcore recommended that service affecting conditions be indicated to the end user on the terminal, or be stored for retrieval on demand.

T-1/PRI Frame Structure

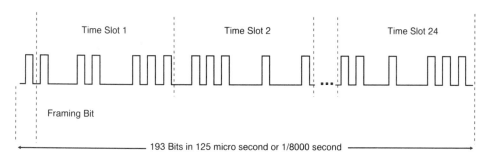

☐ Represents one PCM sample from each of 24 voice channels
☐ Framing bit follows fixed pattern over 12 frames, or over 24 frames with EOC and CRC
☐ Rigid format facilitates switching on PSTN

In-service bit rate is 1.544 Mbit/s +/-4.6 ppm, worst case; normally synchronized to stratum 1 clock, +/-0.00001 ppm; during maintenance, rate may vary +/- 32 ppm.

Figure 2-16

Primary Rate Interface (23B or 30B + D)

Compared to the basic rate interface (BRI), the Primary Rate Interface is relatively simple. PRI has the same pulse shape, framing, rate, and other electrical characteristics at the U, T, and S reference points. Even the R interface seen by older terminal equipment could be a "plain" T-1 sharing these layer 1 characteristics (for the physical/electrical interface).

A DS-1 Primary Rate Interface (PRI) is divided into TDM channels using standard T-1 frames *(Fig. 2-16)*. The pulse shape is the same as that defined for the T-1 or the DSX-1, the digital cross-connect found most often in central offices. This shape is essentially a square wave, nonreturn to zero (NRZ, or 100% duty cycle) pulse, with a peak value of 2.4 to 3.45 V at the transmitter. Some 20% overshoot on the leading and trailing edges is tolerated within defined limits.

PRI requires the Extended Superframe (ESF), which has an embedded operations channel (EOC) and a CRC-6 for error checking (two more TDM channels) in the framing bits. The older D4 superframe doesn't have these features. The EOC carries alarm notifications, statistics, and error indications.

On a T-1, time slot 24 is the D channel, if there is one present on the interface. A signaling messge on the D channel of one T-1 can control a call that passes through up to 19 other T-1 interfaces. That is how an H_{11} channel of 1.536 Mbit/s is managed.

On an E-1, signaling messages use TS-15 (actually the sixteenth time slot, as they are numbered 0 to 31). This is the same TS occupied by ABCD signaling bits when that form of signaling is used on an E-1. The first time slot (TS-0) carries framing codes and a small amount of other overhead.

Traditional T-1 did not allow more than 15 zeroes in a row. While 1s are sent as pulses (of alternating polarity) 0s are represented by no pulses sent. The receiver of a string of 0s has to count bit time intervals with no clocking information from the sender. This can soon lead to errors. Voice meets the ones density requirement by never generating an all-0s byte; there are no long intervals without a clock reference. Data, however, may contain long strings of 0s that are meaningful and must be transmitted faithfully.

You will find different methods to allow unrestricted user data in a time slot.

- **T-1:** the transmitter substitutes a code word for any all-0s octet. Binary 8-zero substitution (B8ZS) changes 00000000 to either $0\ 0\ 0 + - 0 - +$ or $0\ 0\ 0 - + 0 + -$. The polarity for the first pulse in the substituted byte is made the same as the last data pulse, creating the first of two bipolar violations (BPVs). The receiver recognizes these BPVs in the known pattern and restores the 8 zeroes.

- **E-1:** Line coding includes a scrambling step to avoid long strings of zeros.

- **Serial interface:** V.35 and other synchronous interfaces have separate leads for clocking and so can deliver any number of zeroes in a row.

What's different between the reference points (U, T, S in *Fig. 2-17*) is defined above layer 1, in the functional groups between those points. At the physical layer, time slot 24 (the last DS-0 channel

PRI Reference Points

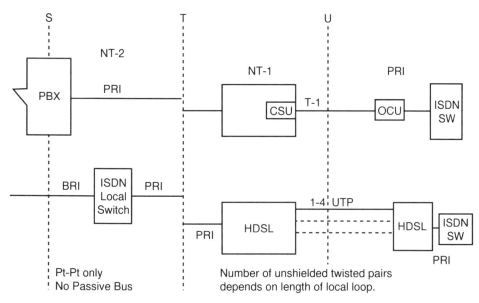

Figure 2-17

in the frame) is the same as the other 23, but it is dedicated to the D (data) channel used for signaling. Clock rate for the line signal at S and T must be extracted from the received signal at the U interface. This requirement arises from the time division nature of the transmission—the network switch must operate at the same speed in both directions. And there is no way to adapt slower rate data to fit the B channel except, in some cases, from the R to the S interface.

Unlike the BRI, where NT-1 extracts receive clock and generates transmit clock, "turning the clock around" at a PRI necessarily becomes the responsibility of the NT-2. The terminal equipment, if it has a T-1 interface, is also loop-timed, so it too sends at the same bit rate it receives.

NT-1 on a PRI is transparent to the data, clock, and framing. It acts as a repeater, not a controller, for timing.

On equipment that has more than one PRI to the same network, the recommendation is to derive clock from the physical interface that carries the D channel from the CPE to that network.

This improves the chance that signaling will be transmitted successfully.

The connector at the demark is an RJ-48, 8-pole modular jack. The terminal side equipment presents an RJ-48 plug on a cable, which itself may plug into the terminal. Pinout is given in *Fig. 2-18*.

Powering the PRI also differs from the BRI or, more precisely, may differ when defined. At this writing, no power is to be applied to the signal leads at the S or T interfaces. The NT-1 at the U interface must not apply power to the loop, but the CO may arrange with the customer to deliver power or loop sealing current.

High-speed Digital Subscriber Loop

The assumption in the standards is that the PRI is provisioned over a traditional T-1 or E-1 transmission system. The T-1 transmitter and receiver in most CPE is designed to the DSX-1 specification. Since DSX was created for use within a central office, the distance limit is about 1250 ft.

T-1 loops normally terminate in a channel service unit (CSU) which acts as a regenerator of the "digital" pulses and performs loopbacks for maintenance. CSUs reliably send and receive data pulses over a distance of about 1 mile (1.6 km) on one twisted pair for each direction (dual simplex transmission). Local loops often are much longer, in which case they need powered repeaters in the outside plant. T-1 is relatively intolerant of wire gauge changes, bridged taps, and sloppy splicing. Pairs must be selected for quality or specially engineered. This proves expensive and slow.

In practice these days, a technology called High-speed Digital Subscriber Loop (HDSL) is increasingly likely to be deployed. Eventually it should replace the traditional T-1 and E-1 lines. HDSL may use the same line coding as the BRI 'U' interface, 2B1Q, but at higher rates. More sensitive receivers (than in T-1 equipment) further improve performance. There are forms of HDSL besides 2B1Q that are available, for example Carrierless Amplitude and Phase modulation (CAP, a modem-like signal). HDSL-2 needs only a single pair.

PRI Interface Modular Connector Pinout

TIA-568A for ISDN Interfaces

Pin[1]	Color[2]	Function		
1	Green	T2	Receive from Network	(NT —> TE)
2	Green/White	R2	Receive from Network	(NT —> TE)
3	Orange/White	R3	Optional shield continuity[4] for pair on pins 1/2	
4	Blue/White	R1	Transmit to Network	(TE —> NT)
5	Blue	T1	Transmit to Network	(TE —> NT)
6	Orange	T3	Optional shield continuity[4] for pair on pins 4/5	
7	Brown	T4	– Optional Power[3]	
8	Brown/White	R4	+ " (this lead grounded if power not floating)	

[1] Pins numbered left to right looking into the jack cavity with locking tab down.
[2] Dominant color/stripe color; twisted pairs may have complementary colors (B/W, W/B) or solid color on one strand, same color with white stripe on the other.
[3] Power transfer may have many configurations; see Index.
[4] Shield continuity is maintained within the premises, particularly on extension cords, but there is no shield continuity across the U interface.

TIA-568B swaps pair 2 with pair 3, changing only the color of the wires on the pins. Electrical performance is the same.

Figure 2-18

2B1Q encoding can carry 1.544 Mbit/s for a mile or more on a single pair (full duplex), rather than the 2 pairs for T-1. Reducing the baud rate from 772,000 (1.5 Mbit/s) to 386,000 (half a T-1) more than doubles the maximum loop length. At 260 kbaud (1/3 a T-1) the distance can double again, reaching more than 5 miles over plain cable, without repeaters. As explained under BRI, 2B1Q transmissions also tolerate gage changes and bridged taps. Consequently, HDSL is much easier and less expensive to install than T-1.

Currently shipping HDSL equipment may have one to four twisted pairs in the local loop. Each pair carries a 0.5 or 0.75 Mbit/s stream of user data, plus some overhead. At both ends of multiple wire pairs, DS-0s are mapped onto time slots of the PRI U interface, in a T-1 or E-1 frame, after removing the additional overhead from each link. LT and NT-1 see only the DSX-1 framing. The HDSL is transparent.

The same sort of HDSL technology is also sold as a short-haul "T-1 modem" or line driver, over a single pair. This version may be useful at the T and U reference points.

NT and CPE at the PRI

Acronym overload aside (see Appendix A), it is time to look at the user side of the local loop. The functions between the U interface and the user may be in one device or several (NT-1, NT-2, TA, terminal). The network sees only what happens at U, so that is where the specifications are concentrated. The devices at the user's site are called, collectively, the CPE (customer premises equipment, *Fig. 2-19*).

The terminal, say an analog phone or non-ISDN digital phone, goes off-hook and dials a number. The CPE interprets the dialed digits and converts them into the correct message format to send to the network over the D channel. Most of the analog to digital conversion will take place in a terminal adapter (TA) function for an analog phone. The remainder of the interworking could be relatively simple, done in a combined NT-1/NT-2, or perhaps a PBX.

At a PRI, recall, the S and T formats are the same as the U format and NT-1 is transparent to clocking.

For most users, the important events happen at R, where existing equipment attaches. The network cares only about U. What happens between R and U is simply 'magic' performed by the CPE vendors.

Physical/Electrical Interface

NT-1 must receive bits at U with an accuracy of one error in 10,000,000 bits, or better. The required layer 1 functions for an NT-1 are available in chips from many vendors of merchant semiconductors.

Because of attenuation along the local loop of up to 16.5 dB (and

'Magic' at the PRI

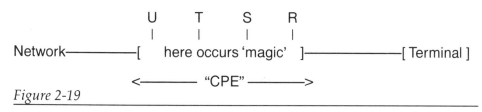

Figure 2-19

another 1.5 dB in an extension cord) the receiver in the NT-1 must detect pulses that are much smaller than the nominal 3 V sent. To ensure reliability, they are tested to 16.5 dB (with an objective of 18 dB) below 2.25 V.

S and T requirements are less stringent. Within a given premises, the attenuation over the cable should be relatively small. For those cases, the DSX-1 specification may apply, with its limitation of about 1250 ft. between devices (half that distance from each device to any cross-connect panel). However, the objective is to reach 3000 ft, which is close to the capability of a CSU. At this writing, Bellcore makes B8ZS only an objective, so S and T could use AMI encoding (without B8ZS).

Standards impose many additional requirements that influence the design of hardware. That is, you must assume that any CPE (NT) certified to meet standards has a certain electrical longitudinal balance, tolerance to jitter and wander, and can handle phase transients. Since the user can have no control over these factors they are not covered here.

Signaling Procedures

The NT-1 and -2 portions of CPE have some housework to do when connected and powered up for the first time:

1. NT-1 synchronizes with the network signal, finds frame alignment, and provides the path for the next step.

2. CPE issues an initialization request to start up the data link on the D channel. An exchange of messages lets the Stored Program Controlled Switch (SPCS), for example the central office switch, give the Terminal Endpoint Identifiers (TEIs) needed by the CPE for addresses; the number of TEIs varies but usually is one per B channel.

3. The SPCS then sends a service profile identification (SPID) to the CPE; again the number may vary from carrier to carrier.

4. The terminal equipment responds with user and terminal identifications (USID/TID).

5. The switch confirms with a message telling the terminals they are attached; for a phone, the message says in effect "you are attached and onhook but not in service."

6. Key sets or ISDN phones with multiple buttons can send a Selected Call Appearance (SCA) to the switch to indicate how the switch should handle incoming calls on each B channel.

7. With that the terminal is ready, so it sends an IN SERVICE message to the switch, which responds with the same message, and the phone is operational.

Note that there may be more than one facility (ISDN PRI line) under the control of one D channel. The second PRI has 1.536 Mbit/s available for an H_{11} channel. If there is no redundancy for the D channel, the limit is two PRIs per D channel. With redundancy from a backup D channel on another facility, non-Facility Associated Signaling on one D channel may extend over 478 B channels (20 PRIs).

The more interesting part of the PRI CPE is in the NT-2 "function group," in the software that handles call control procedures. There are vendors of ISDN software who license the source code to various hardware vendors. Thus the NT products from different manufacturers may be running the same software for signaling and other higher layer functions (management, for example). Common software should improve compatibility.

NT-2 software must distinguish among the various modes, speeds, and bearer services *(Fig. 2-20)* when they are requested in a call setup message. Voice and audio (modem) services are particularly affected, in different ways.

The public switched telephone network (PSTN) delivers many types of information in the form of call progress tones, recorded announcements, and intercept messages. Even when ISDN signaling has an equivalent D channel message, the user may need

PRI Modes and Bearer Services

Mode	Circuit	Packet
Later 3 Protocol	Null	X.25
Later 2 Protocol	Null	LAPB,LAPD
Later 1 Protocol	µLaw, V.110	Null (ESF, B8ZS)
Information Transfer	Speech 3.1 kHz audio Unrestricted digital information	
Bit Rate	64 kbit/s, Nx64 kbit/s 56 kbit/s*	
Structure	8 khz Octet Integrity	
Establishment	Semi-permanent, on-demand	
Configuration	Point-to-Point	
Symmetry	Bidirectional (full duplex)	

Figure 2-20   *56 kbit/s is rate-adapted to 64 kbit/s via V.110 format

to have the audible form. For example, a modem can detect and act on an audible busy signal. Attached to an analog port on the NT/TA, however, a modem will never see the digital message for a busy signal (and wouldn't know what to make of it anyway). The NT itself could generate the busy tone for the modem, but the network delivers the information more easily by telling the CPE to cut through the B channel to the modem while the switch sends digitally encoded tone. The CPE easily converts any bit stream to an audible form, without interpretation.

Likewise the ISDN switch in the central office must be configured to anticipate the need to deliver audible information as well as the digital messages. This explains in part the complexity of ordering ISDN lines. On a pure data line, the ISDN switch might simply send the disconnect message with cause #17, user busy. But for a voice or audio line it would first send a progress message to tell the NT that "further call progress information may be available inband" or "inband information or appropriate pattern now available," and put the busy tone on the B channel.

More detailed information is found in Chapter 3.

Broadband ISDN (ATM or Cell Relay)

ISDN was created as a circuit-switched service. To reach greater bandwidth than a PRI, the functionality of the ISDN was duplicated to a great extent on the cell-switched services known as ATM (asynchronous transfer mode) or cell relay. That is, the ability of a user to connect a channel on an ISDN access line to a channel on any other subscriber's access line has a parallel in the switched virtual circuits (SVCs) of ATM. A sharp difference exists between the narrow-band ISDN (N-ISDN, which is BRI or PRI) and so-called broadband-ISDN (ATM):

- N-ISDN is based on time division multiplexing (TDM) within the voice PSTN. Each channel or connection reserves and consumes a constant amount of bandwidth as long as that connection is in place. Some form of idle signal fills the channel when the user has no information to send. The connection consists of dedicated time slots on transmission links and a full-time cross-connect in each switch.

- ATM's cell-based "virtual" connection (VC) is based on an overlay network of ATM switches that may share transmission lines with the PSTN. A VC consumes bandwidth only when the user of the connection is sending information. The idle condition of a VC usually is an absence of cells to or from that user. When a VC is idle, the same transmission line or channel may carry cells for other users on other VCs.

TDM and VC structures and operations are entirely different, but the delivered services and functions are very similar. Therefore, by extension of the BRI and PRI ISDN concepts to higher speeds, the ATM services were originally called Broadband ISDN (B-ISDN). However, ATM took on its own identity and is no longer much associated with ISDN.

Cell relay was supposed bring ATM services closer to the original ISDN ideal of "all things to all people" than the circuit-switched network ever could be. ATM promised huge bandwidth to each desktop—some equipment offering 155 Mbit/s from an individual workstation.

But funny things happened to B-ISDN on the way to the market:

- ATM took on its own identity;

- Ethernet took over desktop connectivity by providing Fast (100 Mbit/s), Gigabit (1000 Mbit/s), and 10 Gigabit LAN speeds.

- the Internet became the touchstone for practically every-thing.

Upshot: ISDN remains narrowband, circuit switched, and more popular than ever.

(For more information on ATM, see the author's *ATM Users Guide.*)

3.

ISDN Network Functions and Features

ISDN is primarily a collection of services provided by the network. ISDN is much more complex than simple transmission (like Switched 56 or DDS services). However, the network can provide advanced services only to terminal devices that can utilize them. At installation, you must ensure that there is a match between the stored program controlled switch (SPCS) in the ISDN network and the customer premises equipment (CPE). Both must be configured appropriately to match each other. This operation might not be trivial. The complete details of configuration options are needed only by designers of the SPCS and CPE. For practical network operations, the following overview and examples should be sufficient to understand the process. With any luck, you will find this chapter and the chapter on ordering ISDN lines are more than you ever need to know.

Bearer Services (Modes)

A brief listing of the bearer modes appeared with the description of the PRI in Chapter 2. Modes apply to the BRI as well. A more complete description of each mode, from the network's point of view, follows. Of the modes, Speech, Audio, 64 K data, 56 K data, and wider channels (H_0, H_1) are required (they are in the switch software now, if not commonly available to end users). Packet service (X.25) is highly developed in most of the world, but in the US the Bellcore specification required only "nailed up" or permanent channels from CPE to the packet handling facility (PHF) or packet switch. It is a future requirement to offer switched access to packet services.

Likewise, frame relay is not yet fully integrated into the ISDN (even though it originated as an ISDN bearer service). The frame switch is treated as a user of the ISDN, taking calls from frame relay access devices (FRADs). The FRADs then send frames to the switch, which relays (forwards) them to another destination. The FR network switch may support switched virtual circuit (SVC) service, which allows VCs to be set up on demand from the FRAD. Permanent virtual circuits (PVCs) are provisioned by the carrier when the customer orders the circuit and stays in place indefinitely. In the case where the FR switch is another ISDN "user," the FRAD sends its frame relay connection request (SVC setup message) in-band, over the ISDN B channel. The switch recognizes a signaling message because it will use the DLCI dedicated to signaling (the same on all FR ports). This two step call request process will be simplified in the future when SS7 interconnects the ISDN and FR networks.

Circuit Speech

ISDN's "killer" application for some time was telephony. Caller ID and other enhanced features take the service far beyond POTS, but Pulse Code Modulation (PCM) in a DS-0 survives. The customer's equipment is expected to apply a 25+ year-old standard, PCM encoding, to analog voice signals received from analog phones. "Native" ISDN terminals use the same PCM format. This means that calls will be compatible between any mix of new ISDN terminals and all the installed switches, channel banks, and analog phones in the world. Backward compatibility is essential for an economically feasible migration to digital technology.

There is no plan at present to use 7.5 kHz audio in the U.S. This higher-quality voice has been defined, based on ADPCM at 64 kbit/s, for many years. So far users have elected to save bandwidth per voice channel rather than improve the sound quality. On ISDN voice terminals, Speech mode is used to originate all voice calls and to receive (terminate) voice calls from ISDN terminal equipment. For complete interoperability, the ISDN equipment that adapts analog equipment to digital services must also be able to pass through those in-band call progress tones and recorded messages generated by the public switched telephone network (PSTN). This requirement applies to the 3.1 kHz audio mode as well.

Transmission equipment within telephone companies may treat speech calls in ways that violate bit integrity:

- On long lines, echo cancellation is standard.

- Signaling information in the form of robbed bits (ABCD bits) may be inserted into the DS-0 bit stream to convert from SS7 to CAS or robbed bit signaling—without affecting voice quality noticeably.

- On some routes, the carrier may apply voice compression, using ADPCM for example, or digital speech interpolation (silence suppression).

- Certain features designed for voice, like call hold, are not supported by data terminals.

Any of these processes would destroy a data stream. Most of them impair modem signals. So there are other services (modes) for data.

3.1 kHz Audio

When a modem originates a switched voice connection, any echo canceller in the path is supposed to recognize the modem sound and turn itself off for that channel. The ISDN 3.1 kHz audio service avoids echo cancellation entirely. Likewise, this mode does not apply robbed bit signaling or voice compression. Audio mode will interwork with channel banks, analog switches, and any non-ISDN equipment that originates a call. This mode is required if there is any non-ISDN portion in the connection path.

The result is an excellent transport mechanism for modems, including Group III facsimile machines. The network presents all calls from non-ISDN terminals as Audio Mode. In the US, voice encoding for Speech as well as Audio modes is always done with μ-Law PCM. The network will convert to A-Law (for an E-1) when necessary. The difference is a simple change in the companding (compression/expanding) table that converts analog voltages (loudness) to a digital number.

Unrestricted Digital Information

The full 64 kbit/s channel, unharmed by any voice processes, is the transparent digital pipe. It is pure transport, capable of carrying any form of synchronous information.

Thinking of ISDN in simple terms, unrestricted digital information is the model of simplicity for a connection. Push bits in one end and shortly thereafter they come out the other end. Bit patterns are delivered exactly as sent—unless there are transmission errors. Typical applications promoted around 1990 were LAN or terminal data and Group IV facsimile, neither of which ever caught on. The new "killer applications" are indeed for data:

- switched Internet access, usually from a router (the author's home has BRI for voice and Internet access, terminated by a router with integral NT-1 and two analog voice ports) and

- backup of other transmission facilities (Equipment introduced in 1999 provides dial backup for frame relay services). A way to inverse multiplex several connections, with statistical multiplexing, is defined in ITU-T recommendation V.120 as well as via BONDING, Multilink PPP, and Multilink frame relay.

CPE may use the byte-sync capability of an NT to ensure that one complete character occupies each time slot in each S/T frame.

However, this is not required and the terminal devices may establish their own framing through the 64 kbit/s channel. On this service it is possible to build innumerable other services. For example, unrestricted digital service is ideal for a router to access the Internet, because the call is usually to a local Internet service provider, thoughe many enterprises maintain Remote Access Servers designed specifically to receive multiple calls from ISDN TAs or analog modems.

Rate-Adapted 56 kbit/s Information

Switched 56 service was popular in the early 1990s. Those installed lines will be around for a while. Economics has pushed new users onto ISDN basic rate access. To provide interoperability, ISDN adapts a 56 kbit/s bit stream to fit in a DS-0. The method is defined in V.110 (or I.463, which is simply a reference to V.110). In effect, the ISDN TA sets bit 8 in each byte to 1. The other seven positions are taken from seven consecutive bits in the 56 kbit/s stream. Note that if the user is sending byte-oriented data, the content of a byte in the ISDN time slot is an arbitrary seven bits from one or two characters.

By themselves, these seven bits have no meaning other than "user information." The terminal devices at each end must set up their own framing to be able to interpret the bits as data. However since each ISDN byte has at least that one fixed "1" (which generates a pulse on a T-1 line) it is possible to send any number of consecutive user 0s in the 56 kbit/s stream.

Whatever device provides the interworking function between the ISDN and Sw56 services (like a CO switch) may have to operate at both clock speeds. Only a small amount of buffering (less than a byte) is necessary to ensure the 56 K and 64 K sides remain synchronized because they still take clock from the same source (the switch). Neither transmitter is ever starved for the next bit. In some cases the 56 kbit/s stream will arrive already rate adapted in a DS-0, as when it comes from a T-1 line created by a channel bank with an Office Channel Unit–Data Port (OCU-DP).

H Channels (Nx64K)

The call setup request may contain an information element with a 3-byte map of the 24 time slots in the PRI. Each bit corresponds to one time slot. Those slots (DS-0s) to be included in the channel are marked 1, other time slots are 0.

Any number of TSs, contiguous or not, may be grouped into an Nx64 channel. Between the NT and the LT, the group of DS-0s is easily interpreted as a single channel—they appear in the same order at both ends of the line, by definition. Inside the ISDN network, however, transmission may be more complex (though hidden from the user and his NT): the DS-0s must be kept in order, even though individually dialed DS-0 calls could be routed over paths with different delays. The ISDN network should find one path for all DS-0s in an H channel to maintain equal delay on all time slots and thus keep bits in order where delivered. Without this care in transporting the DS-0s as a group, the TEs at both ends would have to resort to some form of inverse multiplexing to ensure bit order within an H channel.

Packet Mode Digital Information

Packet mode, based on X.25 packet format, is intrinsically different from the circuit switched modes of ISDN. The packet handler (or packet switch) is assumed to be part of the ISDN switch itself, not an external function like the current frame relay service. X.25 is not much used in the US.

Rather than request a connection to a specific network address for packet mode service, the ISDN terminal sends a call setup request for packet mode service. The address is assumed to be the integral packet handling facility (PHF).

Once it gets a packet mode connection (on a TDM circuit) to the PHF, the TE sends to the PHF a connection request with a specific address—the X.121 address of the called X.25 entity. The request produces an SVC connection within the 'X.25' network. Recall that X.25 is an access method to a network, and does not define its internal structure or the format used between packet switches.

The X.25 call request over an ISDN may contain specifics for most of the parameters in *Fig. 3-1*.

X.25 Parameters
Supported by PAD Function in Terminal Adapter or Terminal Equipment

Set at Service Subscription
One-way channels
Nonstandard default packet size
Nonstandard default window size
Default throughput class
Local charging prevention
Bar incoming/outgoing calls
IEC preselection
Reverse charges accepted (NI-2)
Fast select accepted (NI-2)

Negotiated per Call
Flow control
Throughput class
Transit delay limit
User testing
Closed user group services
Reverse charging
RPOA selection
Fast select
Basic business group

Futures
Service based on Network User ID
Call deflection
OSI (protocol) network services (NI-2)

Figure 3-1

Flow Control

The packet assembler-disassembler (PAD) and the switch (PHF) have default values for the maximum size of a frame and for window size (the maximum number of frames that may be sent before the receiver must acknowledge correct receipt). If either device requests a different value (longer or shorter packets, more than the default of 2 for a window size) the other responds with an acceptance or with another value. The nature of the negotiation is that if the requested size is not acceptable, the reply will ask for a smaller, but not a larger, value. TE may be configured to accept a range of values from the network, or reject a call request (by clearing the call) if a request is not met.

In the future, packet mode TE may have to accept an unconditional notification (requested value in a call setup request, for example) or clear the call.

Throughput

Total volume of data intended to be transmitted is indicated by a bit rate request. This parameter may be limited by the carrier to a value selected (and paid for) when the service was ordered.

Transit Delay

Certain types of connections are sensitive to total delay. To ensure that an SNA session does not time out, for example, a terminal would request a maximum delay (specified in milliseconds) via the Transit Delay Selection and Indication feature. The PHF will return an estimate of the trip delay over the path to the selected destination.

Optionally, the packet terminals (or PADs) can exchange information in the signaling messages (End to End Transit Delay Negotiation) that estimates the round trip delay without the aid of the network. The requesting end sends the maximum desired time (in ms) in the call setup request; the called end adds processing time at its site and returns that value in the call connected response. These values may be set manually or automatically.

If the network is not able to provide assurance of meeting the limit, the call may be cleared.

User Testing

Besides an internal loopback, the packet TE must be able to:

- place a call to itself, to see its own call request message.
- call the network's 'echo station' which will return all messages of a bit error test.

Closed User Group

The user selects a group address from a list or index. The TE translates the selection into an additional feature in a call request message that selects a closed user group (CUG). While a CUG usually limits communications to other members of the group, ISDN offers options of additional inward only (receive calls from outside the CUG) or outward only access (place calls outside the CUG).

Reverse Charging

The PAD may request the network to charge the called directory number. The PAD may be configured to accept such calls, or to reject them all.

Recognized Private Operating Agency Selection

RPOA is the equivalent of an interexchange carrier for packet data. Users may pre-subscribe, or indicate an RPOA in a call request by including its Data Network Identification Code (DNIC).

Fast Select

Any signaling message allows the caller to send up to 16 bytes of user data. Fast Select increases the payload potential to 128 bytes. A caller may restrict the called party to clearing the call request, or may allow the called party to accept the request and set up a regular packet connection. Interpreting the data is up to an application, not the PAD itself.

Basic Business Group

A user group whose members can dial each other with abbreviated addresses (less than the full X.121/E.164 format). All users are on a single switch, so the carrier can limit the total resources consumed by programming the switch. That's why it's called a stored program controlled switch (SPCS).

Network User Identification

An address attached to the user rather than the line. NUI can be a billing account that allows mobile users to access a network without incurring charges against the DN at the calling line (future).

Call Deflection

Sends a call to another DN by indicating that DN in a call clear response. A TA or terminals uses this feature to forward calls.

OSI Network Services

Similar to the parameters for X.25, but for OSI protocols.

Standards Define Functions

The CCITT started the ISDN process with a large number of "Recommendations" issued from the early 1970s onward. Specifications were fairly complete in the 1988 recommendations (Blue Book). Some of the key documents are listed in the bibliography. The successor organization, the International Telecommunications Union, Telecom Standardization Sector (ITU-TSS), has continued the work.

As was then common, the parties participating in the drafting of the final documents had their own interests to protect. Typically, the result was not a single, unequivocal specification, but a framework with multiple options. In many Recommendations, each special interest has its own option.

In practice, each switch vendor, country, or carrier chose to implement its own option. Result: nobody matched. A switch from vendor A would not work with a switch from vendor B, nor even with the same customer premises equipment. Incompatibility was one of the major forces that retarded deployment of ISDN in the US.

Europe started to get beyond this stage in the late 1980s. Partly, the European Commission imposed common requirements on all member countries. If a piece of CPE was acceptable in any one country, it was supposed to be acceptable in all countries. Today these specifications are known as "Net 3" (BRI), "Net 5" (PRI), and "Net 33" (ISDN telephones). They set common technical requirements (CTR) for radio frequency interference and susceptibility, static discharge tolerance, network protocol compatibility, and so on. CTR approval is good for all countries of the EC, though each country may issue a separate approval number. However, as late as 1995, passing a test in one country could win approval in a second country even though the CPE wouldn't actually work there. Naturally, the French are unique.

US National ISDN

Finally in the US, the various "flavors" of ISDN came together in the mid-1990s into a single version that lets switches of different makes interoperate. Bellcore (with work continued by Telcordia) and the National Institute of Standards and Technology (NIST, formerly the National Bureau of Standards), working with the North American ISDN Users' Group defined a single flavor: National ISDN (NI). It is based on a specific set of choices from the ranges of values and selections allowed by the ITU-TS standards. Because all change is gradual, the introduction of NI was planned in stages. *Fig. 3-2* contains a summary of features in the first stage, NI-1.

NI-1 is an implementation agreement which expresses a consensus on what options are to be used and which are to be excluded because they are not yet supported uniformly. Insisting on additional features would make systems incompatible. NI-1 is a subset of the full ITU-TS functionality. For example, in NI-1 you are limited to one or two terminals per S interface on a BRI. A single terminal may access both B channels and the D channel. If there are two terminals, each uses one B channel and they share the D channel. ITU allows up to 8 terminals per BRI, any of which may use either B channel, or both simultaneously. Further, an NI-1 terminal needs a separate directory number for each type of connection (voice, unrestricted data, packet mode, etc.).

Activation of a terminal, and particularly deactivation to conserve power, is defined by ITU but not part of NI-1. The recommended power source for NI-1 terminals is local commercial power (wall transformer). For NI-2, the recommendation is changed to reflect the international preference for powering TE over the S/T interface, from the NT-1, using the signal leads as power leads (power supply 1, see Chapter 5). Activation of terminals is part of NI-2.

The most widely deployed makes of ISDN central office switches are AT&T, Northern Telecom, Siemens/Stromberg-Carlson, and Fujitsu.

- AT&T (now Lucent) was earliest in the US, and has the most switches installed. Consequently it has more switches with "custom ISDN," which is grandfathered but no longer available for new installations. It started deploying NI-1 in 1994.

- Northern Telecom (now Nortel) had started to conform to NI-1 in the early 1990s. Its 1995 software generic for the DMS switch supported NI-1. Nortel's proprietary version, in service at some sites, was no longer being offered by carriers to customers for new installations after 1995.

- Siemens/S-C has caught up with NI-1.

National ISDN-1 Features

For Voice and Circuit-Switched Data

Call Forwarding
    Variable, Busy, No Answer
    Reminder notification
    Redirecting number, reason
    Courtesy call
Automatic Callback, Intra-switch
Call Hold And Retrieve
Additional Call Offering
    Unrestricted, busy limit
Flexible Calling
    Consultation hold, Add on
    Three, Six-party conference
    Conference hold and retrieve
    Add held call to conference
    Implicit, Explicit transfer
Calling Party Number Identification
    CPN privacy
    Network-provided number delivery
    Redirecting Number, Reason
Message Waiting Indicator
Display Services
Electronic Key Telephone System
    Multiple DNs per terminal
    Multiple DN appearances
    Hold, Retrieve
    DN bridging
    Intercom calling
    Multiline hunt group
    Automatic, Manual bridged call
        exclusion
    Abbreviated, Delayed ringing
    Call Appearance call handling
    Analog members in EKTS group
Station Message Detail Recording
Multiline Hunt Groups
    Linear, Circular, Uniform hunting
    Stop hunt, Make busy
    Analog members in hunt group
Basic Business Group
    Simulated facility groups
    Include non-ISDN lines
    Restricted access (various)
    Deny Originating, Terminating

Business Group Dial Features
    BG dialing plan
    Intercom, Attendant access
    1-digit, Abbreviated dialing
    Dial access to PL, ARS
    Code restrictions, Diversion
    Direct Outward, Inward dialing
Call Pick Up
Access To All Existing Analog Features

For Packet-Mode Service

ISDN Call Control
    D channel packet on BRI
    Provisioned B channel packet (BRI)
Packet Call Control
    CUG, additional in, out dial
    Reverse charging, Acceptance
    RPOA (IXC) selection
    Preselection of single IXC
    DTE packet binary parameters
    X.75 and some X.75 utilities
    Automatic message accounting
        (except aggregate records)
User-to-User Call Control
    Fast select, Acceptance
    16 bytes of data in call request
X.25 Supplementary Services
    1-way logical channels (in/out)
    Calls barred (in/out)
    Flow control, Throughput negotia-
        tion
    Default throughput (up to 9.6K)
    Non-standard defaults for packet
        size, window size
    Transit delay
    User testing
Basic Business Group includes packet
    Packet in BBG dialing plan, hunt
        group
    Non-hunt DNs for hunt terminals
    In-band calling number ID

Figure 3-2

NI-2 Features Beyond NI-1

Uniform BRI Configuration

Restrictions relaxed from NI-1
  Up to 8 terminals supported
  Any terminal uses any channel
  Single DN for all terminal modes
Common features per Bellcore Generic
Requirements
  Electronic Key Telephone Service
  Call forwarding
  Call hold and retrieve with B channel
    reservation
  ACO for circuit mode calls
  Flexible calling (conferences,
    transfers)
  Calling number ID services (delivery
    of net-provided CN, redirecting
    number and reason, number
    privacy, etc.)
  ISDN Display Service
  Call redirection reason not required
CAUSE information not uniform in IE

Enhanced Data Capabilities

  Dialed B Channel Packet on BRI
  X.75 packet net interworking
  Prevention of X.25 reverse charging

PRI Switching and Signaling

Call Control per Full Standard (TR1268)
  Voice, Data, Packet modes
  Multiple DS-1s controlled by D
    channel
  D Channel backup
  Per-call access to services in
    uniform manner (in and out for
    WATS and non-ISDN FX lines or
    tie trunks)
ESF required for PRI layer 1
Only Yellow alarm supported in EOC
Interworking with SS7
Calling Number Identification (delivery
  of net-provided CN, redirecting
  number and reason, number
  privacy, etc.)
Switched DS-1, FT-1 capability

Operations Support

Parameter download from switch for
  automatic terminal configuration
Minimum of 15 critical features in
  uniform manner (out of 154 items
  in NI-2, up from 129 items in NI-1)

Figure 3-3

Migration to National ISDN 2 and Beyond

The definitions of National ISDN-2 are completed. They offer additional services and features *(Fig. 3-3)*, but are planned to be backward compatible with customer premises equipment built to the NI-1 standard. Older CPE built to NI-1 will continue to work when moved to a switch that offers NI-2, or when the switch is upgraded. However, the functionality of the NI-1 terminal will remain at the NI-1 level.

Major work has been done on the PRI definitions in NI-2. These are broader (adding switched DS-1 and switched FT-1), more uniform in offering services.

CPE for proprietary and 'custom ISDN' may not fare as well. While no carrier is likely to kick a user off an existing "custom" service, moving older ISDN access devices to new locations, giving them new directory numbers, or adding features may not be possible. Local exchange carriers apparently plan to migrate away from the older proprietary formats as a way to simplify support and management while improving compatibility. Older installations may be grandfathered for some years (perhaps not permanently), but carriers will not offer new connections of that "flavor." The suppliers of the hardware may be able to upgrade some of it with new software, at least to be compatible with NI-1.

There is also the possibility that changes in the switch software will reveal some incompatibility in the CPE, preventing it from operating even at its old feature level. A hardware feature called FLASH memory promises the least troublesome fix. FLASH memory is a form of non-volatile programmable read-only memory (PROM) that accepts and stores new information while the chip remains installed in the equipment. Until the early 1990's most CPE held computer programs in firmware (PROMs) that could not be rewritten while in place; they had to be removed and erased by exposure to strong ultraviolet light. Bug fixes and new features required physical replacement of the chips with new ones holding the latest program code.

With downloadable FLASH memory, you can send the new program code to a remote device over the network. Depending on how much FLASH memory is installed, the device may continue to operate normally during downloading or it may have to be taken out of service for a few minutes. It is not necessary to visit the site, open the case, or handle parts.

FLASH is the most promising upgrade path to NI2. There will be incentives to upgrade CPE as the ISDN network in the future will offer many additional functions, certainly some that are not anticipated.

Impact on Existing CPE

Because the ISDN access link (the local loop from customer site to central office) is digital, all analog traffic (voice, fax, and modem) must be digitized by the customer's equipment. In most cases this requires some new hardware. Two types of vendors see this as a major opportunity:

1. Makers of equipment with integral ISDN interfaces will suggest you replace your PBX, router, or whatever you have, with a new one. Certainly that will do the job, and allow you to utilize all the new ISDN features, but it may not be necessary.

2. A second group of vendors will offer many forms of "terminal adapters" to put in front of (or inside) your customer premises equipment (CPE). TAs convert analog signals to digital format. They let existing CPE continue to function—over ISDN services. But TAs can't grant access to ISDN features that the older CPE cannot support—like calling party number display on 2500 phone sets. Perhaps the TA could offer this service by having its own display. Not being integrated into your main equipment, the less expensive TAs are unlikely to offer much in remote management from centralized workstations.

Eventually, all CPE hardware will have integral ISDN interfaces, at least as an option. Until then, you may hang on to what you have and adapt it to ISDN.

Common Channel Signaling

An analog local loop has only one vehicle to convey voice, signaling, everything: the loop current. When you pick up an analog phone (go off-hook) you close a switch to draw d.c. current, telling the switch you want service. Rotary dialing a digit interrupts that current a number of times equal to the dialed digit. Voice conversations, modems, and DTMF dialing tones modulate the loop current. All this happens, necessarily, on the same wire pair (channel). This signaling is "channel associated signaling" (CAS).

When interoffice trunks were digitized (with channel banks) the concept of CAS was retained. Robbed bit signaling changes a 0 to a 1 to indicate a phone is off-hook, the bits are toggled to emulate pulse dialing, and so on. These bits are within the voice channel DS-0: channel associated.

ISDN not only converts everything to digital form, it separates conversations and user data from signaling, putting them into separate channels. On the access link (BRI or PRI) the channels are created by time division multiplexing. Only one channel is used for signaling, and is common to all connections: hence "common channel signaling" (CCS).

Rather than flip individual bit values, CCS employs messages, gaining the power to offer new services through flexibility in the message format. With CCS, signaling was no longer limited to 12 (or 16) DTMF tones, the few multifrequency (MF) tones, or the 16 possible values of 4 robbed signaling bits on a T-1 line.

Inside the ISDN this separation of signaling from voice goes further. Not only is signaling outside the voice or bearer channels, it is on a physically separate network *(Fig. 3-4)*. The first common channel interoffice signaling (CCIS) was Signaling System 6 (SS6) introduced into the Bell System in 1976. It was a packet-switched network that carried information and coordinated call setup among central office switches. Line speeds were very slow, often 2400 or 4800 bit/s among the packet switches (Signal Transfer Points, STPs) and between an STP and a voice switch.

When SS6 was introduced, almost all phone subscribers had analog lines. SS6 never much affected users. The major exceptions were "phone phreaks" who had learned to manipulate analog tone signaling. Their tricks didn't work with CCS because they had no way to reach the digital domain of STPs from the parallel PSTN. Phone companies no longer had much need for their clumsy eavesdropping on suspected phreaks. Standardization efforts resulted in SS7 recommendations from CCITT in the 1980, 1984, and 1988 publication cycles. Speed and flexibility of signaling were increased each time. After wide deployment of SS7, and the extension of CCS to CPE via D channels, SS7 remains internal to the telephone companies (they interconnect their systems with each other and a very few major customers). So far, average customers can't reach SS7 directly, but you may need to understand something of how it works.

SS7 Inside the ISDN

When originally deployed, Signaling System 7 was built on packet switching equipment separate from the voice circuit switches. Transmission (historically a different telco department from

CCS Network Parallels PSTN

SS7 makes possible sophisticated services, faster connections.

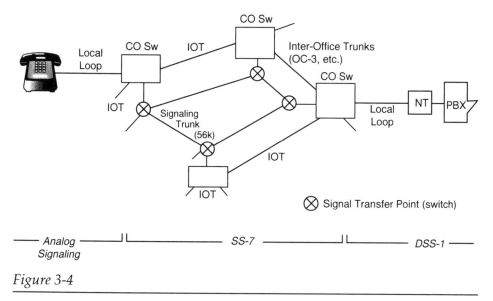

Figure 3-4

switching) probably carried SS7 without knowing it—just another DS-0 channel. As large switches have evolved, they brought the signal transfer point function inside. SS7 and voice may share the same physical switch, but they remain logically separate.

Compared to SS6, SS7 has higher capacity (minimum of 56 kbit/s trunks between STPs) and more features defined through specific signaling messages. The complete specifications for SS7 consist of dozens of documents totaling thousands of pages. Fortunately, because it is hidden within the telcos, you don't necessarily need more than a broad understanding of how SS7 works. The relationships among the protocols and ITU recommendations are shown in *Fig. 3-5*.

SS7 was developed about the same time as the OSI 7-layer model for a protocol and roughly follows that paradigm. Not all functions of the OSI network layer were considered necessary for the Telephone User Part (TUP), the original protocol that carried trunk switching commands. Only a portion of OSI layer 3 was included in layer 3 of the Message Transfer Protocol (MTP-3).

Signaling System 7 Protocol Stacks

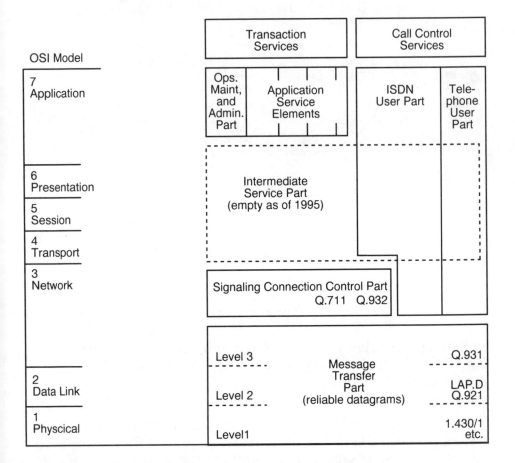

Figure 3-5

The balance was later added in the Signaling Connection Control Part (SCCP).

Though still used in some parts of the world, TUP has been succeeded in N.A. by the newer ISDN User Part (ISUP) for all matters related to call control. ISUP offers more options than TUP, and supports many of the features popular with ISDN, like caller ID, user-to-user signaling, and virtual private networks. An example of ISUP for call set up between switches is in *Fig. 3-6*.

Underlying SS7 are relatively conventional layers 1 and 2.

ISUP Messages for Call Setup

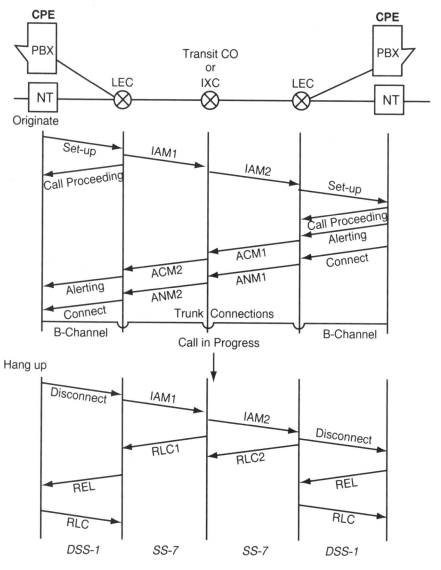

Figure 3-6

1. Message Transfer Part, Physical Layer (MTP1) is full duplex, with a speed of 64 kbit/s assumed for design purposes. Actual speeds range from below 10 kbit/s to a full T-1.

2. MTP-2 Data Link Layer is a typical HDLC frame, with a 2-part address, data, and error check fields. MTP-2 ensures accurate delivery of message signal units (MSUs) between two adjacent devices. The format for MSUs is in *Fig. 3-7*.

Note that the bit order of transmission is left to right, starting at the top, like English text. This statement will be repeated where appropriate because some of the documents in the bibliography use various other layouts to show frame structure.

Multiple MTP-2 links are often installed in sets between two points for redundancy and added reliability. A destination point code (DPC) indicates one of these link sets. The value in the Signaling Link Selection (SLS) field of the message header points the message at a particular link. DPC and SLS form the address for MTP-2 frames. In the event of a link failure, messages continue on parallel links in the set. If the process that receives MSUs experiences congestion it can exercise flow control by shutting off individual links temporarily. MTP-3 recognizes flow control, and routes messages to other links in the set by manipulating the value of SLS. Users of SCCP are identified by SubSystem Numbers (SSNs). They may be in either section of the upper layer (ISUP or TCAP).

MTP-3 has only one destination per link set. SCCP, above it, may have to send information to other SCCP entities at many locations, not just adjacent devices. So SCCP needs to do routing, which may require translation of a directory number (received from higher up) to a DPC and SSN at the other site (or the next transit node). A service control point (SCP) may be invoked over SS7 to complete the number translation.

Service provided by SCCP can be either connectionless (CNLS) or connection oriented (CO). CNLS works well during failures that require alternate routing. Time and complexity to reconfigure PVCs diminishes the potential that the C-O service will ever be deployed for signaling.

Message Signaling Unit
Transfer Part - 2

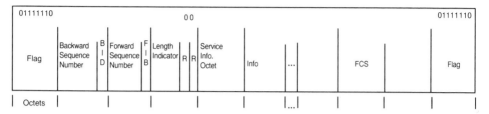

Figure 3-7

When a longer message is segmented to fit the 255-byte limit of MTP-2, SCCP tells MTP-3 to make the SLS the same for all segments. This ensures all segments follow the same path and cannot get out of order (except rarely, during failure transients).

The Transactions Capabilities Application Part (TCAP) handles signaling that is not associated with call control; for example, credit card validating and 800 number translation. TCAP may simply forward a unidirectional message from an application. TCAP can also dynamically set up "dialogues" for continuing transactions of multiple messages. All messages in a dialogue have the same identifying number. A dialogue is very much like a virtual connection between users of TCAP at different locations.

Note that SS7 is between switches, or between a switch and an SCP, etc. All links are within the telephone company. What the user sees on the D channel at the BRI or PRI is different.

D Channel Access

What the NT and user's CPE sees is D channel access signaling, technically known as Digital Subscriber Signaling system 1 (DSS1). It is defined by the Q.900 series of recommendations from ITU-T. Translating DSS1 to SS7 is the subject of Q.699. DSS1 allows the CPE to communicate with the stored program controlled switch (SPCS) that provides the ISDN service. The SPCS does not pass DSS1 messages to the SS7 packet switch (STP). Rather, the PSTN switch digests the DSS1 message, interprets the content, and if necessary may generate a SS7 message to another switch or information source. SS7 messages usually are sent to a nearby STP for forwarding.

There is a component of a DSS-1 message that can be forwarded from the caller to the far end terminal: the user-to-user information element. Users may also exchange data over the D channel in X.25 format, if supported by the carriers at both ends of the connection.

In August 1995 AT&T announced that properly equipped AT&T PBXs which have ISDN access to 5ESS switches can communicate with each other over the D channel. This link allows proprietary features like "call park" and message delivery on LCD displays to work between sites as if every user were on the same PBX. The transport mechanism is user-to-user data in a signaling frame, using the Q.931 format. This capability was defined in the 1989 edition of AT&T's custom PRI specification.

DSS1 Signaling Example: SETUP

As an example of how DSS1 signaling operates, consider a fundamental message, the call SETUP request *(Fig. 3-8)*. This is the message that the terminal or CPE sends to the network to request a new connection or call. SETUP is a good example that contains most of the features found in any signaling message:

- basic data link (layer 2) frame format;
- error checking and correction;
- variances in each direction, user-to-network and network-to-user (there are also some differences possible between BRI and PRI);
- many information elements (IEs), some optional. As noted, transmission order is read like text, left to right. Octets may be marked with "msb" and "lsb" to indicate most and least significant bit locations within the octet.

LAP-D: Layer 2 Header Fields

Like all DSS1 messages, SETUP is framed in HDLC. That is, the bit stream in the D channel contains repeating "flags" (01111110) when there is no message to send. The HDLC hardware inserts a 0 after five consecutive 1s so the flag never appears within a message. The receiver removes every 0 following five 1s between flags. The frame consists of a layer 2 header, a layer 3 header, information elements (IEs) that constitute the message content, and an error checking code (ECC). Figures in this book are arranged to indicate transmission order when read like English text. Within an octet, the least significant bit is sent first, though the most significant bit may go first in some types of transmissions and in certain fields. For example, even in a LAP-D frame, the frame check sequence at the end is sent MSB first.

The particular form of layer 2 used here is Link Access Procedure for the D channel (LAP-D or LAPD, Q.921). It's first byte is a flag, followed by 2 bytes of address, two bytes of control information, information elements that make up the message, and a frame check.

The 2-octet address consists of:

Address Extension. The first bit of each address byte indicates "extension," or whether there is another byte following. It is 0 in all but the last octet, 1 in the last octet of the field. The same technique is used in other fields as well.

Command/Response Bit (C/R) is carried transparently through the network, and is not part of the address. It may be acted on by the terminal. Some polled protocols, among others, use the 'command' indication when the address is that of the receiver, and the 'response' indication when the sender puts its own address in the field. In a D channel, both ends are in a point-to-point relationship and use the same address (the DLCI, see next item). The value of C/R is 0 when the user (TE) issues a command or the net responds; C/R = 1, when the net issues a command or the user responds.

Data Link Connection Identifier The other 13 bits in the address are called, collectively, the DLCI. SAPI may be considered the most significant bits of the DLCI; TEI, the least significant.

Service Access Point Indicator (SAPI) routes a packet to a particular protocol or function. Value 0 is signaling; those frames go to SCCP (or TUP). 16 indicates data, to be routed to SCCP and a higher level protocol. Management messages use SAPI=63. Other values in the 6-bit field are reserved or available for national definition.

DSS1 SETUP Request

(1)

Layer 2 Xmit Order	Layer 3	Bit Value	1	2	3	4	5	6	7	8 / msb	Length (Bytes)
Flag			0	1	1	1	1	1	1	0	1
Data Link Address			lsb	1/0 EA	C/R	SAPI	SAPI	SAPI	SAPI	0	2
			1 EA	TEI							
Control (info. Frame)			1	N(S)						0	2
			1/0 p/f	N(R)						0	
	Protocol Discriminator		0	0	0	0	1	0	0	0 / Q.931/I.451	1
	call Reference		0	0	1	0	Length			0	
			Call Ref. Value							0/1 flag	3 (from SPCS)
			Call Ref. Value								2-3 (from SPCS)
	Message Type		1	0	1	0	Setup	0	0	0	1
	Bearer Capability		1	1	1	0	IEID	0	0	0	4-9
	Info. Xfer Capability		0	0	0	1	Length (bytes to follow)	0	0	1 ext.	
			0	0	0	1	0	0 Q.931	0	1	
	Xfer Mode and Rate		0	0	0	64 K Circuit mode	1	0	0	1	
	Rate Multiplier		1	1	0	6 x 64K	0	0	0	1	
	Protocol, Layer 1 Rate		1	0	0	Rate Adaption	0	0	0	1/0 ext.	
			1	1	1	56 K per DS-0	1	0	0	1 ext.	
	Protocol, Layer 2		0	1	1	LAPB (X.25)	0	0	1	1	
	Protocol, Layer 3		0	1	1	X.25, packet level	0	0	1	1	

(2)

Layer 3			1 / lsb	2	3	4	5	6	7	8 / msb	Bytes
Channel Identification			0	0	0	1	IEID	1	0	0	3-8
			0	1	0	length	0	0	0	0	
			Channel Selection	D ch?	Prefer/Exclsv	Spare	P/BRI	I/F ID	1 Ext.		
			Interface Identifier							1	
			1	1	Ch. Type	0	Map	1	0 CCIT	1	
Channel of 6 DS-0s Mapped on PRI			0	0	0	0	0	0	0	0 24D	
			17	1	1	1	0	0	0	0	
			9	1	0	0	0	0	0	16	
			1	0	0	0	0	0	1	8	
Calling Party Number			0	0	1	0	IEID	1	1	0	10-14
			1	0	0	0	length	0	0	0	
			1	0	0	National Number, ISDN numbering plan	0	0	1/0 ext.		
			0	0	0	User provided, presentation allowed	0	0	0	1	
			IA5 encoded digits							0	
			Up to 14 octets								
Calling Party Subaddress			0	0	1	0	IEID	1	1	0	4-23
			1	0	1	0	Length	0	0	0	
			0	0	0	spare	1/0 od/ev	Type: user spec	0	1 ext.	
			subaddress								
			of up to 20 octets								

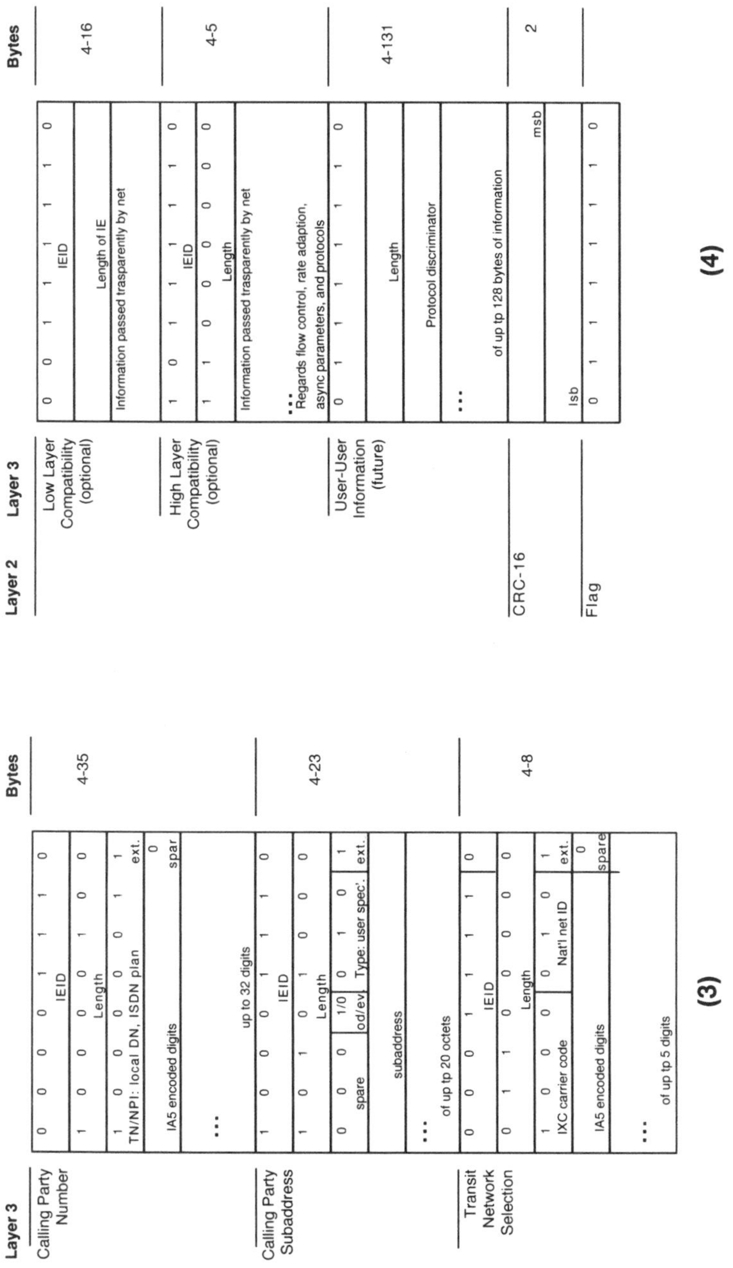

(4)

		Bytes
Layer 2	Layer 3	
	Low Layer Compatibility (optional)	4-16
	High Layer Compatibility (optional)	4-5
	User-User Information (future)	4-131
CRC-16		2
Flag		

Low Layer Compatibility: IEID 0 0 1 1 1 1 1 0, Length of IE, Information passed transparently by net

High Layer Compatibility: IEID 1 0 1 1 1 1 1 0, Length 1 1 0 0 0 0 0 0, Information passed transparently by net

Regards flow control, rate adaption, async parameters, and protocols

User-User Information: IEID 0 1 1 1 1 1 1 0, Length, Protocol discriminator, of up tp 128 bytes of information, msb, lsb

Flag: 0 1 1 1 1 1 1 0

(3)

Layer 3	Bytes
Calling Party Number	4-35
Calling Party Subaddress	4-23
Transit Network Selection	4-8

Calling Party Number: IEID 0 0 0 0 1 1 1 0, Length 1 0 0 0 1 0 0 0, TN/NPI: local DN, ISDN plan 0 1 0 1, ext. 1, spar 0, IA5 encoded digits, up to 32 digits

Calling Party Subaddress: IEID 1 0 0 1 1 1 1 0, Length 1 0 1 0 0 0 0 0, spare 0 0 0, 1/0, od/ev, Type: user spec., ext., subaddress, of up to 20 octets

Transit Network Selection: IEID 0 0 0 1 1 1 1 0, Length 0 1 1 0 0 0 0 0, IXC carrier code 1 0 0 0, Nat'l net ID 0, ext. 1, spare 0, IA5 encoded digits, of up tp 5 digits

Figure 3-8

Terminal Endpoint Identifier (TEI) marks each logical session or virtual connection within a SAP. The user may assign a TEI to a PVC, at subscription; the range of 0-63 indicates this type. Most TEIs are assigned automatically (by the layer management function in the network), during the call setup procedure, from the range of 64-126. Broadcasts use TEI 127 (all seven bits = 1). For comparison, a Frame Relay frame uses the "DLCI" the address. However, the Frame Relay Forum implementation agreement dedicates 3 of the 13 bits to specific functions (1 for discard eligibility and 2 for explicit congestion notification). That leaves 10 bits in the FR DLCI. It is not divided logically into SAPI and TEI.

Control Field occupies 2 octets after the address. Three formats are defined:

1. Information (I), in which the control field carries two sequence numbers:

   - N_S, Number Sent, incremented by the sender for each frame sent. The number cycles through 0127 (modulo 128). The receiver tracks the sequence and acknowledges each number to detect loss of frames and request retransmission of lost frames.

   - N_R, Number Received, tells the other end what NS is expected in the next I frame, which acknowledges frame NR–1. I frames carry signaling messages, ensuring error-free delivery.

2. Unnumbered Information (UI) frames have no sequence numbers and provides only "best effort" delivery service. The error correction missing from UE is supplied by a higher-layer protocol.

3. Supervisory (S) frames carry sequence numbers that acknowledge frames when there are no I frames flowing. S frames may also impose flow control by halting transmission on a DLCI (receiver not ready message, RNR message) and later starting it again (receiver ready, RR message). LAP-D procedures to set up a data link, acknowledge frames, request retransmission of lost or errored frames, and so on are essentially the same as HDLC. A more detailed discussion is beyond the scope of this book, but is covered in the author's *Frames, Packets, and Cells in Broadband Networking*.

Standardized DSS-1 Message Types

As Documented by Bellcore;
National Authorities and Carriers Define Additional Messages

Interface Rate			
BRI	**PRI**	**Call-Control Message**	**Meaning or Function**
x	x	ALERTING	'Ring back': called party is being alerted
x	x	CALL PROCeeding	Sw: request OK, call initiated
			TE: please wait
x	x	CONNect	Call has been answered
x	x	CONNect ACKnowledge	Optional symmetry for TE; Sw uses after CONN
x	x	DISConnect	Call ending
x		HOLD	Frees B channel without ending call
x		HOLD ACKnowledge	Confirmation that HOLD has happened
x		HOLD REJect	Failure of HOLD request
x		INFOrmation	Supplementary keypad digits or other data
x		NOTIFY	Sw sends TE data regarding special service
x	x	PROGress	In-band audible progress tones
x	x	RELease	Starts call clearing (hang up)
x	x	RELease COMplete	Starts or continues call clearing
	x	RESTart	Request to put channel or interface in idle mode
	x	RESTart COMplete	Channel or interface now idle
x		RETrieve	TE returns to a held call
x		RETrieve ACKnowledge	Sw confirms that held call is reconnected to B ch.
x		RETrieve REJect	Sw refuses (or can't) RETrieve held call
x	x	SETUP	Initiates call between TE and Sw
x		SETUP ACKnowledge	To TE, Sw needs more info to set up call
	x	SERVice	Request to change signaling to another D channel
	x	SERVice ACKnowledge	Change to another D channel complete
x	x	STATUS	TE: responds to Sw ENQuiry with call state
			Sw: protocol error detected, and cause
x	x	STATUS ENQuiry	Sw requests STATUS from TE

Figure 3-9

Q.931: Layer 3 Information Elements

Layer 3 consists of Information Elements (IEs) assembled into
different Message Types *(Fig. 3-9).* Every message starts with a
1-byte protocol discriminator that identifies the format as Q.931

(also defined in TA-NWT-001268 and other documents). Q.931 is the same for BRI and PRI, though some information elements my not be used at both interfaces or may take different forms at each. All messages also contain a call reference IE and an IE that identifies the message type. A message may end there (CONNect ACKnowledge) or have many more IEs (SETUP).

Just so nobody runs out of codes for IEs, there is a special IE (SHIFT) that substitutes another "code table" with entirely different meanings for the same value in the TYPE field. In the US, there is only one IE in the national codeset (Codeset 5) for operator access. The "Custom" versions of ISDN offered previously by each switch maker have additional information elements. AT&T's, for example, occupy Codeset 6.

The first three IEs are not part of a Codeset because they are interpreted before the first possible appearance of a SHIFT IE.

As seen in SETUP, IEs are constructed similarly, with an identification *(Fig. 3-10)*, a length indication (1 byte encoded with the binary number of octets in the IE to follow after the two mentioned so far), and content or information fields. IEs appear in the same order in all messages (increasing binary value in the TYPE field). This allows the receiver to check for missing IEs. Here we start with the three IEs in every message type, then go alphabetically.

Some IEs defined in Q.931 may be futures, not supported immediately by every switch. Even in NI-2 some IEs are not expected to be handled exactly the same by all switches.

Protocol Discriminator

For D channel signaling this first IE is always 0001 0000, indicating Q.931 is the defining specification.

Call Reference

The originator of a call at the interface assigns a 2-byte number that stays in place until the call ends. Since both sides may assign the same number at the same time to different calls, a FLAG bit (the last one in the first byte) distinguishes a number from the originator (flag=0) and the same number sent back from the called side of the interface (flag=1). The FLAG makes all call reference

DSS-1 Information Elements

Codeset 0, In Effect At Start of Each Message Received

Information Element Identifier Coding

8	7	6	5	4	3	2	1	
1	:	:	:	-	-	-	-	Single-octet information element:
1	0	0	1	0	X	X	X	Locking shift
0	:	:	:	:	:	:	:	Variable-length information elements:
	0	0	0	0	0	0	0	Segmented Message
	0	0	0	0	1	0	0	Bearer Capability
	0	0	0	1	0	0	0	Cause
	0	0	0	1	1	0	1	Extended Facility
	0	0	1	0	1	0	0	Call State
	0	0	1	1	0	0	0	Channel Identification
	0	0	1	1	1	1	0	Progress Indicator
	0	1	0	0	1	1	1	Notification Indicator
	0	1	0	1	1	0	0	Keypad
	0	1	1	0	0	1	0	Information Request
	0	1	1	0	1	0	0	Signal
	0	1	1	1	0	0	0	Feature Activation
	0	1	1	1	0	0	1	Feature Indication
	0	1	1	1	0	1	0	Service Profile Identification
	0	1	1	1	0	1	1	Endpoint Identifier
	1	0	0	0	0	0	0	Information Rate
	1	0	0	0	0	1	0	End-to-End Transit Delay
	1	0	0	0	0	1	1	Transit Delay Selection and Indication
	1	0	0	0	1	0	0	Packet Layer Binary Parameter
	1	0	0	0	1	0	1	Packet Layer Window Size
	1	0	0	0	1	1	0	Packet Size
	1	0	0	0	1	1	1	Closed User Group
	1	0	0	1	0	1	0	Reverse Charging Indication
	1	1	0	1	1	0	0	Calling Party Number
	1	1	0	1	1	0	1	Calling Party Subaddress
	1	1	1	0	0	0	0	Called Party Number
	1	1	1	0	0	0	1	Called Party Subaddress
	1	1	1	0	1	0	0	Redirecting Number
	1	1	1	0	1	1	0	Redirection Number
	1	1	1	1	0	0	0	Transit Network Selection
	1	1	1	1	1	0	0	Low-Layer Compatibility
	1	1	1	1	1	0	1	High-Layer Compatibility
	1	1	1	1	1	1	0	User-User Information

All other values are reserved

Figure 3-10

values unique. All-0s is the global reference, used for restarting a LAP-D link.

Message Type

The seven-bit value is preceded by a 0, reserved for a potential extension bit if the field ever needs to be made larger. The earlier table lists the Types and their values.

Bearer Capability

One of the five bearer modes of NI-1 takes up to 8 octets to indicate because the IE is constructed to be generalized. There are many values reserved for future standardization. Two optional octets, for protocols, are used only with LAP-B/LAP-D (layer 2) and X.25 (layer 3). This IE appears only in the SETUP message.

Call State

A dozen possible conditions exist for a call at any given moment. The Status message reports one of these states for a particular call reference.

Called Party Number

For ISDN calls, the called number is either local, national, or international in the ISDN numbering plan (E.164). The "Type" field may also indicate "unknown number in unknown plan," in which case the SPCS uses it with less interpretation. Dialed digits in a SETUP message are indicated by 7-bit IA5 codes. In addition, the first bit of each octet is an extension bit, marking when the IE ends. Up to 15 digits may be carried in this IE. The size will have to be increased soon to accommodate longer phone numbers.

Called Party Subaddress

Up to 20 octets, each carrying a digit in IA5 code, indicate a sub address within a call to a particular directory number. Both ends of the ISDN connection must be provisioned to allow subnet addressing in the SETUP message.

Calling Party Number

This IE is almost the same format as Called PN, is also used only in SETUP, but has an additional octet, just before the dialed digits start, which indicates:

- if the number was provided by the caller or is the default number saved in the switch (perhaps for a group of trunks);

- whether the called terminal is allowed to display the Caller ID (CPN). Blocking of caller ID is controlled by this IE. The caller's number is sent to the far end, but the terminal is told not to peek.

Calling Party Subaddress

Same format and function as for Called PS.

Cause

As phone equipment gets more computerized (for example, as PBXs open their computational platforms to third-party software developers) the potential to use information from the network rises dramatically. ISDN signaling offers a host of reasons for what just happened, delivered in the CAUSE information element. While an analog phone might reduce every error/cause message to fast busy tone (re-order tone), a computer can distinguish causes by location and specific errors. Rather than say just "oops," the Cause IE tells generally where it happened (private or public network, local or long distance carrier, and which end). Another field offers a broad explanation, indicating one of eight classes from normal operation to protocol error. The last field may contain a more detailed cause code, like "bearer capability not implemented" or "recovery on expiry of timer."

Just to make sure there are enough values available, the standard provides room to designate one of three complete sets of causes. These are ITU standard, national standard, and network specific. There are a lot: the topic covers 10 pages in the Bellcore specification.

This IE may be repeated in a message when there are multiple events and causes to report.

Channel Identification

At a BRI the choice of channels is limited compared to the variety on a PRI. The IE format is the same, but the IE at a BRI ends sooner, in no more than 3 octets (of which two are the IE identifier and length fields). All the information is in 7 bits of the remaining octet [bit position]:

- Interface ID [7], always 0 on a BRI meaning "this line," the same one that carries the D channel; on a PRI, can be 1 to say the channel is not on the PRI carrying the D channel. In that case, the next octet is the interface identifier; if 0 on a PRI (implicit identification), the interface identification field is omitted. At a PRI the interface identifier is present if more than one T-1 is controlled by a single D channel. With 7 bits in the field, one D channel can control up to 125 additional T-1 interfaces. More than a second T-1 is allowed only when there is D channel redundancy.

- Type of interface [6]; 0 is BRI, 1 is PRI.

- One bit [4] says whether the sender of the IE insists on this channel ("exclusive"), or merely prefers it.

- On packet mode calls, a bit [3] indicates if the D channel is acceptable or if only a B channel will do. This bit is always 0 (no D channel) for circuit-mode calls.

- Channel selection on a BRI is simply two bits [1 and 2] where a 1 value in either position indicates that the channel is selected. Both bits set means any BRI channel is acceptable. On a PRI there are two options: if bit [5] = 0, the next byte is a single channel number; if [5] =1, the following 3 bytes are a map of the 24 channels. Setting a bit position to 1 means it is included in the channel, 0 means not included. The bit map allows any group of DS-0s on an interface to be grouped into an Nx64 channel.

High/Low-Layer Compatibility

These are two separate, but related, information elements that pass between end terminals and allow them to check for compatibility. The network does nothing more than check that the maximum lengths (6 and 16 octets) are not exceeded.

Many of the information codes standardized by CCITT (now ITU-T) are antiques: all the current and once-common async bit rates down to 50 bit/s, for example. However some codes will be used more in the future, like the one for Group 4 facsimile (64 kbit/s digital fax). There are fields to set bit rates different in each direction, identify telematic services and the other bearer modes, rate adaption formats, and specific protocols at layers 1, 2, and 3. In all, enough coding to fill 20 pages, but buried in the terminal equipment or not it is usually beyond the user's control.

Progress Indication

This is one of the information elements that causes incompatibility among CO switches. One field codes for location of an event along the call path (public or private network, local at what end or a transit net in the middle). Another field gives the progress description value to notify of a delay, the presence of in-band tone or announcements, or to warn that there is interworking with a non-ISDN network that prevents end-to-end digital transfer of B or D channel information.

To make the IE more interesting, it can represent ITU, national, or network-specific codes and events.

Transit Network Selection

When this IE is absent a long distance call is placed with the default or the presubscribed interLATA carrier. This IE may contain the same access code dialed to select another carrier from "equal access" phones. The 3- or 4-digit carrier ID is coded as IA5 characters, one per octet.

Locking Shift

This IE is different in that it is only 1-octet total. It affects the interpretation of IEs that follow by changing the Codeset to one whose number is carried in a 3-bit field. The new codeset stays in effect until changed again. For example, to reference the operator services IE, a national element not defined by ITU in Codeset 0, the locking shift IE calls out Codeset 5. Then the following IE is interpreted as a national definition.

Non-Locking Shift

Applies to only first IE following. Not supported in National ISDN.

Operator System Access

Because of the standardized format for IEs, this one takes three octets to convey one bit of information: when the customer dials 0, is the operator who answers the default service or one presubscribed?

Packet-Layer Parameters

For packet-mode connections, the caller need not indicate any of the parameters that are preconfigured (on both CPE and the line). The called party, however, must be notified of what parameters the caller would like to use. Thus the network-to-user SETUP message includes the following information elements that do not appear in the user-to-network direction. The parameters are added by the SPCS at the origination point.

Packet Layer Binary Parameters

Boiled down, there are 5 bits to indicate:

- Modulus, 8 or 128;

- Delivery confirmation, link by link or end to end;

- Request for Expedited data; and

- Fast Select treatment (2 bits). This IE is a future requirement, not uniformly implemented in NI-1.

Packet-Layer Window Size

This IE requests a layer 3 window size for each direction of transmission when a call request reaches the called terminal.

Packet Size

Requests a maximum packet (layer 3) size for each direction, when a call request reaches the called terminal.

Closed User Group

The caller may include a closed user group facility in the X.25 Call Request packet sent to the network. The CUG number in binary-coded decimal form is translated to IA5 characters and included in this IE in the SETUP signaling message sent from the ISDN network to the called terminal. Another field indicates if there is outgoing access allowed beyond the CUG.

Information Rate

Four fields indicate the information transfer rate and the minimum rate for each direction. This is a future requirement at this writing.

Reverse Charging

This IE appears only at the termination of a packet mode call request, in SETUP. The originator puts the reverse charge request in the layer 3 X.25 CALL REQUEST packet.

Network Transit Delay

The originator puts a request in the X.25 CALL REQUEST packet, which causes the network to deliver an ISDN SETUP message to the called end with this IE included. The key field is a 16-bit binary number representing the maximum delay in milliseconds.

End-to-End Transit Delay

Values for cumulative transit delay, requested end-to-end transit delay, and minimum end-to-end transit delay are copied from the X.25 INCOMING CALL packet received by the network (SPCS) from the calling terminal.

Features Based On Signaling

The place of SS7 in various telephone network features varies from simply delivering the calling party's number (for CLASS) up to supporting complex transactions among multiple locations (number translations). Historically, that is for the first 15 years after ISDN was defined, telephone companies couldn't get past the wonderful technology of SS7. They were wildly enthusiastic about CCS because it saved big money by preventing toll fraud

and reducing call setup times. Telcos believed in ISDN and continued to convert CO switches to digital with ISDN capability.
The paying public yawned and ordered more POTS lines.

Then about 1994 the sales effort shifted in focus from "ISDN"
(the infrastructure) to appealing, useful features like Caller ID,
Follow Me numbers, distinctive ringing for each member of the
family, voice mail, private numbering plans within companies,
and so on. ISDN became an overnight success. In a classic story
of marketing and technology, people bought the benefits.

CLASS Features via CCS

Among the most frequently purchased services are a wide range
of Customized Local Area Signaling Services (CLASS). These
are offered under various names *(Fig. 3-11)*. Some are available
on analog lines, though the features depend on CCS to deliver
the calling party directory number (CPN) to the terminating
switch. For each service, the switch makes different use of the
number.

Each feature may have its own list of CPNs, for filtering or screening. List lengths may be limited when they are held in the switch
itself, for lack of memory capacity.

Service Control Points

While communications enables some network features, information is necessary for other features. Collecting, storing, and dis-
seminating information is the job of service control points (SCPs).
These nodes are linked to switches, operator stations, and each
other via SS7. For a given call, several switches may ask for information from one SCP, and deliver information to another. All
the combined functions of SS7, SCPs, and the switches lead to
the concept of Advanced Intelligent Network. AIN offers service features not available from the switches alone. The added
value is in software, at SCPs, and may be proprietary to a carrier.
Some companies are strong in billing features, others have something special to offer in cellular roaming.

Customized Local Area Signaling Services

Automatic Callback	Dials directory number of last caller; CCS used to verify that both ends are idle.
Distinctive Ringing	Applies different ringing pattern if caller DN matches customer's list.
Selective Call Forwarding	Customer lists CPNs to be forwarded.
Call Waiting	Screening announces only listed CPNs.
Repeat Dialing	Redial last number called from this station.
Selective Call Rejection	Screens out calls from listed CPNs.
Caller ID	CPN delivered during ringing, may be displayed before answering.
Customer-Originated Trace	Allows called party to send own phone number, CPN, time, and date to phone company office or police.

Figure 3-11

800 service (freephone)

When introduced in 1967, 800 numbers were linked to one carrier by the 3-digit "exchange" (800-NXX). That carrier dedicated a specific customer line to each 800 number, and charged for calls delivered there. The customer couldn't originate calls on those lines. Changing carriers meant changing numbers.

The FCC ordered in 1989 that carriers provide "800 number portability" to allow user to change carriers and keep the same number. To make numbers portable, they first had to be made virtual. That is, a number could not correspond to any particular access line or local loop. The solution is an SCP that translates 800 numbers to directory numbers. Today, when a caller dials an 800 number, that caller's serving office uses SS7 to send a query to an SCP *(Fig. 3-12)*. After consulting its database, the SCP returns the "real" phone number, which the switch uses to forward the call to the pre-subscribed interexchange carrier for delivery to a specific line.

Changing the data base modifies how a call is handled. For example, calls to one 800 number may be directed to different call centers based on time of day. Calls to an office on local holiday may be redirected temporarily to a site that is working that day. The subscriber may switch to another IXC via a change in the data base.

CCS Enables Number Portability

Number Translation Also Underlies Call Forwarding and Follow Me

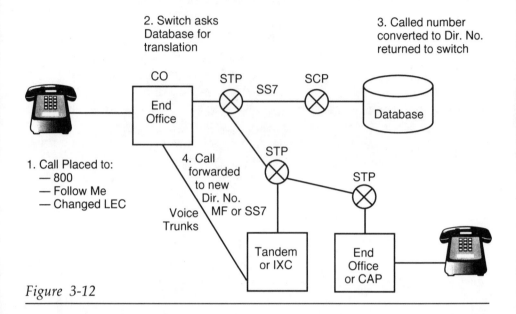

Figure 3-12

Alternate Billing

Credit card calls are one example where automated billing is not applied to the calling phone. The switch redirects the call to an SCP where recorded voice or signal tones prompt the caller to enter a credit card number. After verifying the number, the SCP confirms to the switch that the call may be completed. Afterwards, the switch sends billing information to the SCP, or more likely another specialized SCP.

Voice Processing

Voice Mail in the central office provides recording and playback of messages stored in digital format at an SCP.

Call Completion function is based on voicemail. When a called party is busy, some carriers offer to record a message and attempt to deliver it later—for a fee. Information from the original call attempt is delivered via SS7 to the SCP, enabling it to try the same call later. Without SS7, the caller might have to re-enter the phone number, or credit card number, making the service cumbersome and less likely to be used. Also known as call completion to busy subscribers (CCBS). If the called party has CO voicemail locally, the message could be transferred there and a 'message waiting' indication left.

Spoken Caller ID applies voice recognition technology to verify the identity of a caller and authorize charges. Verification from the SCP travels over SS7 to the switch.

Collect calls are completed without an operator through an SCP. The caller uses a special 800 number that reaches a voice processing SCP. The SCP records the caller's name, which is played back to the called party when asking if the charges will be accepted. Speaker independent voice recognition looks for the "yes" answer, then completes the connection and bills the called party.

Number Translations

The "Follow Me" phone number service relies on essentially the same function as 800 number translation. In this case, the customer gets to change the database directly, and at any time. SS7 links the database to switches, and to the dial-in service node that supports customer changes.

Call forwarding is a number translation, though it may be implemented in the switch itself rather than a separate SCP.

Virtual Private Networks (VPNs) are similar to closed user groups based on the public network. The carrier, through various SCPs including number translation, can offer a private numbering plan, abbreviated dialing, special billing reports, and other services to an organization with many locations. These original VPNs are not related to LAN-based VPNs offered more recently on the Internet.

4.

Ordering ISDN Lines

For a basic phone line (Plain Old Telephone Service or analog POTS) these days you usually don't even get to choose rotary dial or touch tone any more—you get both automatically. ISDN lines are not so simple. There are differences in the proprietary "flavors" of ISDN on the central office switch. Then there are the many options on how the switch provides service features and which features the customer wants (and doesn't want).

Fortunately, most of the difficulty for end users disappeared quietly, invisible compared to the Y2K problem being solved at the same time. Where the worst case used to require the user to pick dozens of parameters to specify ISDN BRI service, the latest ISDN customer premises equipment has a single ISDN order code (IOC) that tells the LEC everthing needed to configure the line. IOCs will be described later in this chapter.

Even if you don't have the manual (and so don't have the IOC) all you need do is tell the phone company which equipment you have (model number, etc.), and perhaps the application, and they will match it to the specific IOC or profile recommended by the hardware vendor.

Your Results May Vary

ISDN access tariffs are all local. ISDN by its nature is local access to the central office: from your serving office your traffic goes on as part of a regular transmission system, almost always over optical fiber in the U.S. What the telephone companies charge for almost identical equipment, features, and functions has an astonishing range. Like POTS, ISDN may have a business rate and a (usually lower) residential rate. Then again, everyone may be charged the business rate, though a residential tariff is offered in many areas.

What happens in the future can't be known, but as predicted in the first edition of this book the trend is toward lower ISDN pricing relative to analog POTS and especially relative to older digital services. In fact, DDS leased lines and Switched 56 circuit-switched channels are generally going up in price. These last two services provide a subset of what is available from a BRI service, yet involve much more manual effort within the telco to install and maintain. For example, most DDS and Switched 56 provides 56,000 bit/s on a 4-wire local loop. BRI provides more than twice that capacity on a 2-wire loop. With easily deployed capability, ISDN BRI continues to be an important offering for LECs facing competition from independent DSL-based CLECs.

For the sake of their own long-term costs, all telcos will have to migrate customers from leased lines and special services (often provided on custom engineered lines) to a generic digital service, ISDN (or frame relay, or ATM, or IP, or something newer). The way to convince users to change services is to show them how to save money for comparable performance or how to increase performance for comparable cost. Thus a BRI, to be attractive, should be less than twice the cost of a POTS line for voice. If so, it would be much cheaper than Switched 56, though this is not always the case as some LECs have charged more to carry an ISDN bit labeled "data" than a bit labeled "voice."

The Federal Communications Commission threw a scare into ISDN fans in early 1995 with a ruling that the subscriber line charge (SLC) be applied to each bearer and data channel, not just the 2-wire or 4-wire "line" that provides the interface to ISDN service. Several RBOCs requested a rescinding of the FCC's order, which was granted, but with reservations that it may be reconsidered in the future.

Originally, the SLC applied to the "line" which, in basic phone service, is a twisted pair of copper wires. That same copper converted to ISDN BRI supports three channels (2B+D). Two pairs are needed for a primary rate interface (PRI, 23B+D). The SLC, $3.50 per residential line or $6 per business line and rising, makes a considerable difference in the economics of ISDN service for most users. For T-1 access the difference is $1728 per year.

The SLC was created by the FCC and levied on customers to replace revenue lost by the local exchange carriers when access charges to long distance companies were reduced. The result was a shift of costs away from long distance callers at the expense of those who didn't make many LD calls.

Configuring For Terminal Parameters

With an IOC, you no longer need plan on serious configuration.

The plug-and-play level of most ISDN products in early 1995 was about the same as add-in cards on IBM PC-AT clones. There were many hardware options to select for the application. Then those same selections had to be reflected in the configuration of the ISDN line from the phone company (see Chapter 3 for that viewpoint).

Help arrived in the form of parameter sets that can be specified with a single designation—an ISDN Ordering Code, described later in this chapter. For National ISDN-1 (NI-1) service the many IOCs represent an advance but still reflect the legacy of "you gotta configure it" from earlier ISDN versions. The real advance comes with NI-2: IOCs are defined, for up to two terminals, in the form of a base package (there is only one) plus up to six "feature modules" for functions like forwarding and voicemail. Details follow the discussion of NI-1 IOCs.

But assuming the worst case, and you must face the full configuration task, here are the factors to deal with. They apply on either the BRI or PRI, though ranges of permitted values could vary; for example, the number of B channels allowed in a connection.

Number of Channels

Under some tariffs, one version of BRI service is offered for D channel access only: the number of B channels is 0. At the other extreme is a PRI with a D channel that controls a second PRI with 24 B channels. That leads to a maximum of 47 B channels possible.

Default: BRI, 2B + D

                PRI, 23B + D

Bearer Services

The ISDN switch is capable of passing a 64 kbit/s bit stream transparently, so why should you have to configure the line for voice or data? Because the "N" in ISDN (the network) may not be transparent.

When a connection request is sent to the network by your terminal equipment, the message specifies what kind of call it is: voice, data, X.25, etc.

- Knowing a connection carries voice, the network might apply echo cancellation to the channel or convert the encoding from A-law (North American version) to mu-law (European version) of PCM. Any of these processes would destroy data, so must be avoided on data connections.

- A data call that is passed to another carrier, or is directed to a subscriber line that has Switched 56 service rather than ISDN service, must be rate adapted to 56 kbit/s.

- When the bearer service is X.25 packet switching, from a Packet Assembler-Disassembler (PAD), the network wants to know in advance what volume of traffic to expect and what specific services the customer wants so the X.25 switch may be prepared in advance.

For calls terminating at your site, the specification of which services are supported allows the switch to reject calls that are incompatible. A request for a packet mode connection makes sense only if the called device can deal with packets. This capability is indicated by configuring the line to support packet mode data when the CPE can receive packets.

Options are: speech, 3.1 kHz audio, circuitmode 64 kbit/s, circuit-mode 56 kbit/s adapted to 64 kbit/s, and packet mode.

Directory Number (ISDN Phone Number)

Each ISDN interface must have at least one DN, but there is no limit on the number that may be assigned. A DN is associated with only one interface. One DN will be the default for the interface.

A basic rate line can have a different phone number for each B channel, or both may be considered a hunt group with a single "directory number." Carriers who anticipate exhausting numbers within an area code will appreciate saving some numbers where they are not needed for additional ISDN B channels.

Your Calling Party Number

Does your equipment have to supply a calling party number with each call request? Default is no, but the line may be configured to require such an information element in the call setup request message. This choice makes sense when many users (many phones) call out on the same interface and you want to identify the individual user (or station) to the network, for billing, determining subscription parameters (like presubscribed IXC), or other purposes.

Screening of the outgoing calling party number, for validity, is performed by the network when CPN is sent. If not valid, the network will deliver the number to the called party anyway, but also adds the default DN for your interface.

You have a separate choice on whether you want your CPN presented to the called device. Default is yes, to permit ANI to function, but the selection can be changed with an information element in the call request.

Subaddress Information

Each user may have another identification, the subaddress information, in addition to the DN. It may be used at the terminating end of a call to route the connection over a private network or to a specific station.

The network wants to know if you plan to send or will accept subaddresses, either yours or the called party's. If you don't want or can't use it, tell the network not to deliver any.

Calling Number Delivery

Do you want to receive the DN of the calling party? Specify yes or no.

Early Cut Through

Normally the network waits for your CPE to confirm that it is ready to receive information (the called extension is off-hook) before opening the channel from the calling party. But when your CPE offers in-band ringing tone while waiting for an extension to answer, you can configure the line to cut through early. This action allows the caller to hear the call progress information.

There is a separate selection of early cutthrough for (1) ISDN and (2) non-ISDN terminal equipment or private network devices behind the NT.

IXC Presubscription

Each interface can designate an interLATA carrier or InterExchange Carrier (IXC). The network will attempt to route circuit mode calls originating here to this carrier. Packet mode calls are handled differently.

Protocol Compatibility Information

End terminals may have special functions or needs. To negotiate compatibility with the equipment called over the ISDN, a TE may send what is called high level and low level compatibility information in the SETUP message. The network does not interpret this information, and can't act on it. However the SPCS that first takes the signaling message will verify the size and format of each information element, including the compatibility information.

Number of Terminals

The S interface point will support a passive bus. That is, the terminal-side transmitter on the NT-2 has parallel connections to as many as 8 TE-1s or terminal adapters. The count limit is imposed in part by the signal reduction as each attached TE or TA soaks up some of the transmission power. Too many devices will reduce the signal strength to the point where none of them gets a usable signal.

Under older ISDN practices, carriers would provision up to four SPIDs per BRI, which are still available under some business line tariffs. National ISDN-1 limited the number of TEs per S interface to two. NI-2 also allows only two terminal configurations on a BRI. For simplicity, and for a lack of "pure" ISDN devices, the number in the US seems unlikely to reach 8, the international standard.

Other local restrictions may apply to "custom" versions of the BRI interface.

Service Profile Identification (SPID)

This is the unique layer-3 identification for each circuit-switched terminal device: phone, fax, computer port, etc. There may be more than one such device associated with an NT-1, NT-2, or B channel. For example, an ISDN TA with two voice ports, a serial data port, and an S/T interface would need unique SPIDs for each port so that someone could dial into them. That is, one SPID, through user configuration of the TA, would identify a fax port, another SPID the data port, and so on.

The SPID is assigned permanently, by the carrier at subscription time, to be unique on a switch. The number may be the 10-digit directory number (DN), if there is only one device or port per B channel, or the DN plus a prefix and/or a suffix. The exact form of the SPID is determined by the carrier and depends on the switch make, the software level (National ISDN stage or earlier software), and local practice.

The National ISDN Council fostered the adoption of a standard SPID format that simplifies assignment. The SPID is the DN (NPA-Nxx-XXXX) plus a four-number suffix to identify a B channel (2 characters) and a specific device on that channel (2 characters). Most equipment then can use a SPID of DN+0101 for most applications.

The SPID traditionally has been programmed manually into the TE and the switch in the CO. The switch uses it when the TE is first installed and initializes the layer-3 link (the TA sends its SPIDs to the switch as a form of identification). Enhancements in NI-2 let an unconfigured TA initialize itself and receive the SPIDs assigned to the line by the LEC—no manual configuration for the end user.

The switch from then on uses the SPID to provide the service (mode) appropriate to that device, even when it shares the DN with another device that needs a different mode. For example, a phone and a router may share a line and a DN, but have different SPIDs to identify which is calling.

Terminal Endpoint Identifier

Each TE must support a layer-2 terminal endpoint identifier (TEI) for each logical connection on the interface. TEI is part of the frame address (2nd octet) at the S/T interface. Individual NT-2 devices may select which frames to capture based on a match of a stored TEI with the frame address. The 7-bit field allows 128 values, but 127 is the broadcast address, always active. Of the remainder, 0 to 63 may be allocated manually, for PVC connections to a frame handler (packet switch).

TEI values from 64 to 126 form a pool from which the CPE and network negotiate a value to associate with each new switched connection they set up.

There must be at least one directory number (DN) for each TEI.

ISDN Ordering Codes (IOCs) for NI-1

The complexity that up until 1994 was the despair of ISDN customers (as well as carriers and hardware vendors) is almost gone. Telcordia (formerly Bellcore), the National ISDN Council, and the National Institute of Standards and Technology (NIST) run a program with NIUF to predefine configuration sets that specify the values of all the possible parameters to be input into the ISDN switch (the "switch translations"). Once defined, these pre-set configurations may be used by anyone.

Generic configuration sets fit broad categories of equipment in typical applications. Vendors can then build default values into their equipment that will interoperate with an ISDN line configured for the same set. It's close to plugand-play, and really there if the TA has an autoSPID feature.

The standard or "Generic" IOCs are widely implemented, replacing those carriers had made up ad hoc. Bell Atlantic, for example, had lumped all the decisions necessary to support the ProShare video conference system under the label "Intel Blue." Rather, Bell Atlantic had three such sets to match the type of switch (AT&T or Northern Telecom) and whether the ISDN "flavor" is proprietary (AT&T Custom) or National ISDN 1. A European version could be yet another set. Now one of the generic IOCs covers almost any need.

Documentation for terminal equipment should state the code (or codes, depending on application) to use. The installer of the equipment need tell the ISDN carrier simply which IOC is wanted. This should be as brief as a few letters, not the individual values for all of the parameters to be set in the switch.

The idea behind IOCs is to select line configuration sets that fit common applications. A given piece of terminal equipment may require different line capabilities (ordered with different codes) for different situations. A hardware vendor might suggest IOC Capability XX for a general purpose terminal adapter, but Capability YY when the same TA is used exclusively for an intense graphics file transfer or LAN interconnection application.

One reason that ISDN equipment had been relatively expensive was that buyers needed extensive technical support to configure and install it. That effort had to be paid for in the price of the hardware, even when the cost to manufacture was modest. As the need for support droped to the level required for a modem, prices for ISDN CPE likewise dropped. A very functional TA/ router with voice ports is only a few hundred dollars (in 2000).

There are three tracks at for defining configurations in terms of an Ordering Code for NI-1. They are distinguished by whether an IOC is associated with a specific vendor's equipment or is intended for a broad application (generic). If generic, there are IOCs that require strict conformaty (one or two letters) and those that require only compatibility (EZ ISDN and a number/letter).

EZ ISDN allows the TA to ignore some features configured on the line, but the TA must not interfer with operation of the switch. This idea is appealing and deserves to take over entirely the specification of ISDN services and the compatibility of CPE.

Generic Ordering Codes for NI-1

Groups of vendors within the NIUF defined core configurations that they saw as fundamental sets of ISDN line features. Each generic core capability registered was proposed to and considered by multiple organizations. There was consensus that the configurations serve a broad purpose—more than to meet the needs of one vendor's equipment.

However, experience in the field with early IOCs determined that some of the original parameter sets are almost never used. These have been relegated to a status called "archived." This means no new lines will be installed based on them, no new CPE will be certified against them, and nobody will support them. These IOCs are marked as such in *Fig. 4-1*.

Each generic "capability" is designated by one or two letters. 'A' through 'Q' were published in the first edition of this book, but more were added. The list may still grow as carriers will support NI-1 for some time.

The Capability represents a fixed configuration for all parameters of line, switch, and service. No variations are supported.

In the mid-1990s, for only $100, you could have bought a Bellcore paperback book (SR-3480) that spells out all the arcane commands to program a CO switch for each IOC, in cookbook fashion, with no background information or even explanations for the acronyms. It takes a book because each brand of switch "translates" these parameters differently. This Bellcore publication also described the process to obtain review, confirmation, and registration of IOCs. Bellcore changed its name to Telcordia when SAIC purchased them from the RBOCs.

Generic NI-1 ISDN Ordering Codes

IOC	Line Set	Interface B	B	D	No. of DNs ATT	NTI	SSC	Features
A	1			p	1	1	1	D-channel packet service only; no CNI
B	3	c			1	1	1	CNI for c only
C	4	vc			1	1	1	
D	6	v		p	1	2	1	Basic D channel packet; no CNI
E	6	v		p	1	2	1	Basic D channel packet; Flex calling for v; ACO(v) Archived
F	6	v		p	1	3	1	Basic D channel packet; Advanced EKTS (v) Archived
G	11	v	c		1	2	1	Flex calling for v: 3-way, hold, drop, transfer; Archived
H	11	v	c		1	3	1	Advanced EKTS (v) Archived
I	14	c	c		1	2	1	no CNI for v
J	15	c	vc		1	2	1	
K	15	c	vc		1	2	1	Flex calling for v; ACO(v)
L	15	c	vc		2	3	1	Archived
M	17	vc	vc		2	2	1	
N	25	c	vc	p	1	3	1	Basic D channel packet; Flex calling for v; ACO(v) Archived
O	25	c	vc	p	2	4	1	Basic D channel packet; Advanced EKTS (v) Archived
P	27	vc	vc	p	2	3	1	Basic D channel packet; ACO(v)
Q	27	vc	vc	p	2	3	1	Basic D channel packet; Advanced EKTS (v); Archived
R		c	c					No voice
S		vc	vc					ANI
T		v	v	p				Archived
U, EZ1		vc	vc					{EZ-ISDN 1} and certain CLASS features
V, EZ1a		vc	vc					{EX-ISDN 1A} and Advanced Call Forwarding
W		c	c	p	3	3	3	Same as R plus D; no voice
X		vc	vc	p	3	3	3	Same as S plus ACO, D, and p
AB		vc	vc	p	3	3	3	Same as U plus all voice featues and p
AC		vc	vc	p	3	3	3	Same as V plus all voice featues and p
EZ-ISDN 2		vc	vc		3	3	3	EKTS with voice features
EZ-ISDN 2A		vc	vc		3	3	3	EKTS with advanced call forwarding
EZ-ISDN 3		vc	vc	p	3	3	3	Same as EZ-ISDN 1 plus p
EZ-ISDN 3A		vc	vc	p	3	3	3	Same as EZ-ISDN 1A plus p

CNI: All IOCs include (calling number identification except as marked.
ACO: Additional call offering   p: B-Channel packet-mode data
v: Circuit-switched voice   c: Circuit-switched data
vc: Alternate voice and data   DN: Directory Number
D: D-channel packet data   *Archived: discontinued*

Figure 4-1

Proprietary IOCs (NI-1)

Specialized terminal equipment may not fit into any of the generic "Capabilities" defined. New applications may call for unanticipated parameter combinations. Hardware vendors have the option of modifying a core or generic capability by making some (not many) changes. The IOC is then designated by the letter of the underlying generic IOC ('X') plus a number ('n'), resulting in the form "Xn."

Changes might be the number of directory numbers (DNs), assignment of a feature key to a location on a smart phone, etc. Changes would not be accepted in this format if they affected the basic interface; e.g., the number of B channels.

ISDN Ordering Codes for NI-2

Ordering codes for National ISDN-2 are very different. There is only one basic function package, augmented by six optional feature modules. Each base package applies to one terminal device; there may be one or two basic packages assigned to a BRI line, which give each terminal access to both B channels. The packages may be different for each terminal.

Unlike NI-1 IOCs with relaxed conformity demands (the EZ-ISDN classes), NI-2 demands full compatibility for certification. That is, the customer premises equipment certified for NI-2 compatibility makes full use of the features for which it is certified. The only exception is for Calling Number Identification, which need not be supported by data-only terminals.

NI-1 IOCs remain valid. Those not archived may be used indefinitely for equipment design and for ordering lines.

Basic NI-2 Package

The minimum set for an NI-2 IOC (called NI2-1) includes one directory number, a terminal equipment identifier (TEI), and one SPID. The terminal may use both B channels if equipped to deal with both at one time. Voice and circuitmode data are included.

With either voice or data incoming calls, the network delivers the calling number and a "redirecting number" (where a call was transfered from) if there is one. This is the equivalent of "caller ID" service on home phones, with added information about a location that forwards or transfers a call using the ISDN network capabilities.

Feature Modules for NI2-1

There are six additional feature sets that can be added individually to either or both basic packages on a BRI line.

Flexible Calling (FC): 3-way conferencing, etc.

Call Forwarding (CF): may operate on ring-no answer or when the called SPID is busy.

Voice Mail (VM): answers calls when terminal equipment doesn't.

Calling Name (CN): the alphabetic information about the owner of the calling number.

Additional Call offering (AC): notice that another call is waiting; the caller hears ringing rather than an immediate busy signal.

Packet data (P): in X.25 format.

To name a specific combination of features modules, their initials (in parenthases, above) are added to "N2": thus the full set is called N2FCCFVMCNACP.

Where can this go? Not much further as ISDN is now mature enough to reduce the pressure for more development. One feature glaringly absent is the frame relay bearer service. Frame relay, recall, started as part of ISDN, then became an offering provisioned almost always over dedicated access like a T-1 or 56 kbit/s leased line. The technology exists to allow a terminal device to dial a circuit-switched connection to a frame relay port on a packet switch, as discussed elswwhere in this book. Switched virtual circuits on the FR service would make the switched access more attractive.

Implications of IOCs

Some time near the middle of 1995, the traffic carried by US phone companies crossed the line from being mostly voice to being mostly data. This watershed event went unnoticed by most people, but those responsible for provisioning data services (including ISDN) have been working at ways to handle the rapidly growing demand for new data connections and changes.

One response has been to offer switched digital service for data (read ISDN) the same way analog service has been offered for voice in the past:

- install the access line permanently to each subscriber premises;

- let the customer provide the phones/terminals;

- let the customer decide how the service is to be used: incoming or outbound; voice calls, data, facsimile, or modem; local, regional, national, or international calling areas; sales, manufacturing, or applications of many kinds; etc.;

- future provisioning actions are limited to turning the service off (and then back on again for a later owner or tenant), and selling highprofit services like call waiting, call forwarding, voice mail, and so on.

Analog voice lines are a generic service. So is ISDN. If users are ever to migrate to digital services via ISDN, the ordering process must be made not much harder than ordering plain old telephone service (POTS). That's what the Capability sets and IOCs do.

Since the Capability definitions have no options, the process cries out for automation. In fact, several RBOCs have started working on a graphical user interface (GUI) for ISDN order entry. The GUI will hide the complexity of setting up a BRI line not only from the customer but also from most of the carrier's own employees.

However, the parallel between POTS and ISDN goes further. Ordering something more complex than POTS and the most common value-added services (usually sold bundled to simplify order taking and provisioning) can be fairly complex. For example:

- a hunt group of analog lines is often installed wrong, may develop problems that are hard to diagnose, and are understood by only a very few people in the average telco business office.

- The same problems and difficulties are more likely for a multidrop data circuit.

- Many consultants make a nice living finding mistakes in telco bills.

Imagine, then, trying to get ISDN service that does not correspond exactly to a specific IOC. At least for a time, using the IOCs means a carrier's ISDN switch is programmed manually from 'cheat sheets' or pages from a cookbook corresponding to each Capability. Explicit instructions for each line parameter lets the person who provisions the service enter the data quickly. That operator probably retains some knowledge of what's happening and should be able to customize if needed.

Jump ahead in time. The order taker, one of many new hires needed to handle a growing volume of ISDN orders, is sitting at a workstation that displays graphical icons to click on. So how will you get anything that is not on the screen? There is no button or icon for what you want, so the operator won't have a way to provision it. Is special assembly the answer? Once in that category, your price may be based on time and materials—in this case the time needed to take your special order, confirm it is a legal configuration supported by the switch, and to input it manually. Could be expensive.

Representatives of regional and national carriers expressed publicly their plan to support all of the IOCs officially registered by Telcordia (Bellcore). However, RBOCs talk openly of charging more to install a line that does not conform to an IOC. They can justify price differences based on their labor costs. Carriers also make it clear that unless Telcordia has tested the translations, and certified them with a particular piece of vendor equipment, they won't even try to troubleshoot a problem—that will be up to the hardware vendor—and you, too, of course.

In 1995, ISDN provisioning was still manual, so any configuration was about the same as any other. Since IOCs became automated, ISDN equipment had to work with a line configuration defined by a registered IOC. For belt and suspenders security, users may want to be sure the specific hardware has been certified by a compatibility lab—trouble shooting support from the carrier may be harder to get without hardware certification. Fortunately, hardware vendors have been building to generic IOCs.

5.

Premises Wiring

The ISDN connection isn't complete until it reaches the terminal, and that usually requires some "inside wiring" on the customer's premises.

Like the local loop, inside wiring is assumed to be unshielded copper twisted pairs. While the ISDN signal will work for some distance over almost any wire that is suitable for analog voice service, better wire allows longer reach between the NT and the terminal.

Garden variety voice wiring has been Category 3 for some years. Earlier, "pre-Category" wire was of unknown characteristics due to both construction and installation practices. Even older wire may be used for BRI installations where ISDN is being added to an existing residential building where such wiring is in place.

Cat. 3 wire, if installed carefully, will carry Ethernet (10BaseT format) at 10 Mbit/s. Cat 3 is adequate for PRI as well as BRI. E-1 interfaces for EuroISDN are moving away from the dual co-axial cable interface to two copper pairs. In new installations, the wire of choice is "Category 5" (or the "enhanced Cat 5" with even longer reach) because it supports emerging technologies like 100 Mbit/s "Fast" Ethernet, Gigabit Ethernet, and ATM at speeds up to 155 Mbit/s. Of course it carries ordinary voice, too.

Wire pairs in a higher category (larger number) are twisted more tightly and more uniformly than lower categories. A shorter pitch between twists is a better match to higher frequency signals and is better at reducing emissions which contribute to signal loss. Consistency in all dimensions gives the wire a nearly constant impedance, minimizing reflections and distortions which occur when the impedance of a transmission line changes along its length.

For the same reason of consistency, higher category connectors must be applied very carefully to cables. Category 5 wire, for example, should remain twisted to within 0.5 inch of the connector contacts. Unravelling the twisted pair for more than 0.5 inch changes the impedance and prevents part of a high-speed digital signal from passing through (energy is reflected back to the source). There will be no effect on analog voice so trying a phone on the new cable is not a good test.

T Interface Wiring

Between NT-1 and NT-2 there can be only one interface: a 4-wire connection consisting of a twisted pair for transmission in each direction (and possibly power being provided over other leads in the 8-wire cable). The T interface is little used in the US and Canada, replaced by chip-level signals on one printed circuit board that contains both NT-1 and NT-2 functions. Outside North America, The T or S interface is the only customer demark point.

This point-to-point configuration *(Fig. 5-1)* calls for a terminating resistor (TR) at each end of each twisted pair of the inside wires. A TR at the end of a line absorbs the power of the signal pulses to minimize reflections back to the transmitter. The terminating resistor can be built into the NT and may have a switch to remove it, electrically, when not needed.

If there are terminating resistors on the cable, they mark the points where the "cable" and NT meet. Practically, the TRs would be

Premises Wiring: T Interface

T interface is point-to-point connection between NT-1 and NT-2

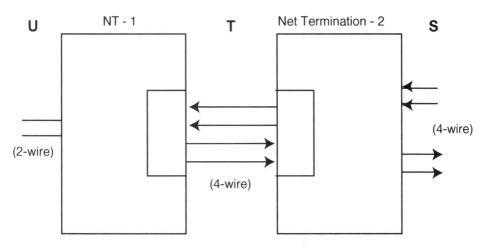

Figure 5-1

mounted on a modular connector (jack) at the interface on the end of the wiring. Many types of NTs have the terminating resistor built in. These TRs may be permanently wired across the connection, or may be removable. Most often a TR is disconnected by a DIP switch or when an internal jumper is removed.

Any cable or cord between the NT and the TR is considered part of the NT. Cords may be permanently attached to an NT or plug into a connector on the NT. If the TR is in the NT, the patch cord is considered part of the wiring, not part of the NT.

ITU standards dictate minimum and maximum lengths for these cords, but only for conformance testing. In practice, you use cords that work, but the testing guidelines are a good rule of thumb for reliable operation. For example, a long extension cord from one TE to a point-to-multipoint S bus could disrupt all the TEs on the bus. Still, it is largely the quality of the wire that determines how long the cable may be in any specific application.

A cord of up to 10 meters connects the interface jack to the TE. Bellcore recommended that all terminals be attached with the same standard cord, to ensure compatibility.

S Interface 'Passive Bus'

In addition to a point-to-point connection between the NT-2 and a single terminal (like a T interface between NT-1 and NT-2), the basic rate S interface supports a point-to-multipoint connection on two twisted pairs. It is called a passive bus because it contains no logical functions (though pulse regeneration or amplification is permitted). Conventionally, wiring diagrams like *Fig. 5-2* may show a single line to indicate both wires in a pair.

Multiple TEs are not allowed on a PRI at the S interface.

The assumption for BRI wiring is that all TEs plug into jacks wired directly on a single run of cable from the NT outward to the TEs. Stubs of up to 1 meter are allowed between the main run wire and the jack for the terminal. Terminal equipment or network terminations plug into these connectors with 4-pair cables (that is, the information is on 4 wires while power may be provided on other leads in a 4-pair cable).

To control electrical characteristics, a terminating resistor (TR) of 100 ohms must be placed across the extreme ends of each twisted pair of signal wires. This TR may be built into the S interface of an NT. That TR may be permanently wired in place, as there should always be one at the NT end of the S bus.

Only one terminating resistor is needed at the far (terminal) end of the S wiring. The TR will be separate (like the terminating resistor on Ethernet coax) if the cable runs past the last attached terminal (the one farthest from the NT). The TR may be built into a TE if it is at the extreme end of the cable. Only one TR should be connected on the S wiring at the TE end. Any TRs in other TEs should be disconnected (usually by removal of an internal jumper or flip of a DIP switch).

Polarity is not critical in point-to-point connections at S or T. But because there is a potential for multiple connections on the S interface, it is recommended that the polarity of the wiring be preserved between NT-2 and any terminal device, even if there is only one TE installed initially. Standardized wiring and modular connectors applied consistently will maintain control of polarity.

Premises Wiring: S Interface Passive Bus

S interface of BRI may be pt-pt, but allows up to 8 devices attached in parallel

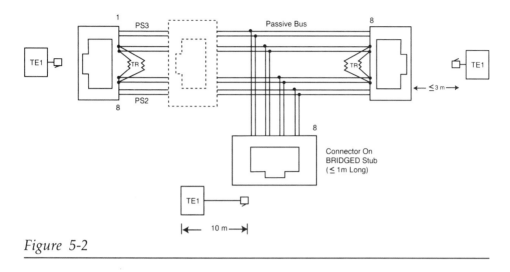

Figure 5-2

As a parallel bus, the wire will accept connections at any point. Both the NT and the terminals may be scattered along the cable. While there is no official limit on the number of TEs permitted on an S interface, for testing the number of TEs is limited to 8. The NT may be attached at any point, but for testing it is at one end, which is commonly the case.

The locations of NT and TEs on the S bus are limited by timing considerations. Specifically, the round trip propagation delay of a signal from the NT to one TE must be within 4 micro seconds of the delay to any other TE. The difference can be expanded if the NT dynamically adapts to different delay times.

Practically speaking, this means the TE connectors on the bus can occupy no more than about 150 feet of cable. Exactly how much depends again on the wire—in this case its impedance, which is a function of wire diameter, insulation thickness, twist pitch, and materials. Most 4-pair Category 5 cable is 100 ohm impedance, which places it between the two examples in CCITT documents:

- 150 ohm impedance cable propagates signals fast enough to cover 200 meters (200 m) within the time limits.

- 75 ohm impedance cable has a lower propagation speed, requiring TEs to be within 100 m of each other.

Commercial 100 ohm cable will fall somewhere between.

Note that it is the round trip signal propagation time from NT to TE that must be about the same for every TE. This says nothing about how large the delay can be. In fact it can be much larger as long as the differences remain small. Conditions resolve into two classes:

1. Short Passive Bus

If all the TEs and the NT are on the same 100–200 m segment of cable, they will meet the propagation time constraints. The quality of the cable in terms of attenuation is not an issue over such a short distance.

However, additional TEs act as stubs on the transmission line and slow down the speed of signals along the cable. To control this effect, the patch cords (4-pair cords with modular plugs) from TEs to the connectors on the passive bus should be kept short.

2. Extended Passive Bus

If the TEs are clustered within 50 m on the cable, the differences in propagation times to the NT will be no more than 2 microseconds. (Note that the round trip delay difference includes two passes over the 50 m, or 100 m of transmission distance.) Since the NT has the hardest problem if round-trip times are very different, when they are nearly the same the NT can be quite far away from the TEs, at least 500 m. Good quality wire should extend the range to 1000 m. The cable length may be limited by attenuation, but more likely by the ability of the NT to deal with longer delay times.

Connectors for Cable and Equipment

The standard connector for ISDN equipment is the 8-position RJ jack. Patch cables have 8-pole RJ plugs at both ends unless wired permanently into the equipment (which is a bad idea as cable will break or develop intermittent errors more easily than equipment). Inside wiring may be terminated with either a jack or

Connector Placement on Modular Cables

4 TWISTED PAIRS

RJ - 45
8 - POLE

Figure 5-3

plug, depending on how TEs and the NT are to be attached.

In the US, the pinout standard for ISDN is TIA568A *(Fig. 5-4)*. Note that it specifies not only which pins are used together on a twisted pair, but also the color of insulation on the wire strand at each pin position. Color matching is necessary to control polarity. Reversed polarity on some ISDN interfaces indicates a test condition that disables normal operation.

The same pinout applies at both ends of an ISDN cable, which is the practice in the data world. This means that a flat cable, untwisted, with an RJ modular plug on both ends, will have the locking tab of the plug on one end up and on the other end down *(Fig. 5-3)*. By contrast, telephony practice is to apply connectors with both lock tabs on the same side of "silver satin" cable—one end's pinout is the reverse order of the other (sometimes called a crossover cable by data people, though that term often means something more specific in terms of the data paths).

Strict cabling and connectorizing procedures have many benefits:

- Cables are always made up the same way, and so are interchangeable.

- Repair to one end of cable (new plug) is made easily without examining the pinout of the opposite end.

- Transitions between wiring methods are standardized (punch down block, modular RJ connector, 50-pin connectors, etc.).

Some documents refer to the twisted pairs by numbers 1 through 4, not color. To preserve polarity, the TIA standard refers to T and R, carryover ideas from analog telephone jacks where Tip and Ring were electrical contacts on the telephone (phono) plugs at the end of switchboard patch cables. T and R can just as well be designated by solid/stripe, or color.

The important consideration is to preserve polarity, and to terminate a given twisted pair in the same pin positions at both ends. Thus minor differences between a manufacturer's literature and the markings on their components are not important as long as you stick with one version for all connections. Markings directly on connectors are the easiest way to remember pinouts.

Be careful when mixing connector brands or product lines to be sure they actually have the same markings—you might have either the A or B variants. Note that some jacks have internal "twists" so the visible arrangement of the wires (the sequence of color codes printed on the jack) may not match how the conductors connect electrically to the positions in the connector.

There are variants A and B to the TIA specification, and connectors are available with either markings *(Fig. 5-4)*. The differences don't matter if applied consistently to all points. However, mixing the two will swap certain pairs between ends and result in line faults.

Higher categories of wire require more precise and consistent connectors. For example, the familiar "66" punchdown block is fine for wiring categories 1 through 3. For category 4 or 5 you may want to ensure maximum distance of reliable transmission by using the newer "110" punchdown block or modular patch panels (RJ connectors and Cat. 5 jumper cables). Extra care is needed in dressing the wires and particularly in the small details of how cables are terminated on jacks and plugs.

To preserve its uniform impedance, Category 5 wire must not be untwisted more than absolutely necessary—no more than 1/2 inch (about 1/2 turn) before it ends on the connector. With plugs this is relatively easy and familiar to anyone who has done phone work.

- The outer sheath of the cable is removed for a short distance.

- The 8 wires are cut off square (all of the same length, the dimension determined from the connector literature).

'S' Interface Modular Connector Pinout
TIA-568A for ISDN Interfaces

Pin[1]	Color[2]		Function
1	Green/White	T2	Power 3 (from TE to other TEs and/or NT)
2	Green	R2	"
3	Orange/White	R3	Transmit to Network (TE —> NT)
4	Blue	R1	Receive from Network (NT —> TE)
5	Blue/White	T1	Receive from Network
6	Orange	T3	Transmit to Network
7	Brown/White	T4	− Power 2 (48 V DC nominal, 32 V min. from NT)[3]
8	Brown	R4	+ " (this lead grounded if power not floating)

Signal pairs should be twisted, Category 3 or better.

[1] Pins numbered left to right when looking into the jack cavity with locking tab down.

[2] Dominant color/stripe color; twisted pairs may have complementary colors (B/W, W/B) or solid color on one strand and same color with white stripe on the other.

[3] Power transfer may have many configurations; see below.

TIA-568B swaps pair 2 with pair 3, changing only the color of the wires on the pins. Electrical performance is the same.

Figure 5-4

- Individual strands (not stripped) are inserted, according to the color scheme, into the plug.

- All contacts are made with a single squeeze in a special crimping tool. Pressure forces each wire in the jack to displace the plastic insulation from the cable strand and make solid contact with the wire. The crimping tool used for 4- and 6-position RJ-11 phone connectors won't work for the 8-pin jacks without a change of dies. A new tool may be required.

High quality jacks from most vendors come in either of two styles: 110 punch down or with a "snap cap" that replaces the punch down tool and remains part of the connector.

The punch down variety tends to be slightly larger, physically, to leave room for the 8-position punch down field. Those parts are joined to the wires in the connector cavity over a small printed circuit board. The 110-style connection properly made with a punch down tool is going to be first class and permanent.

The snap-cap connector (also known as IDC or Insulation Displacement Connector) terminates wires in almost the same way as punch down types, between the prongs of a split post. However the connector is smaller, leaving no room to punch down individual wires—all must be set at one go using the cap of the connector as the tool. The cable end is prepared with slightly more sheath removed, about 3 inches. Each strand is fed through the jack body in its own tunnel, which positions the wire over the split post. After all strands are in place, they are snugged up by pulling on the cut ends (to minimize the untwisted length behind the jack) and the extra wire is cut off and discarded. When the cap snaps into position (most of us need pliers to seat it, some brands have a special tool) it forces each wire into its electrical contact. The cap stays in place on the jack to protect the connectors and also clamps the cable sheath to act as a strain relief.

Another caution is to ensure clear labeling of the connectors (jacks) on all equipment and particularly in wiring closets (jack fields, etc.). The standards call for the same 8-pin modular connector for the U, S, and T interfaces. Some interfaces may have power available on certain pins. Certain vendors use the same 8-position jack for an RS-232 supervisory port. In the same wiring closet with ISDN terminations you may find a T-1 line ending in an RJ-48 connector—the same form factor and pin count. The pinouts of these various interfaces are deliberately different in ways that prevent some of the worst mismatches (T-1 loop current of more than 100 volts applied to an ISDN receiver circuit designed for 2 volts). But it is best to know what functions each connector supports, through clear labels, and to connect accordingly.

Power Distribution

Outside of North America, where the carrier supplies NT-1, the carrier may also supply the power to run it (via loop current as with a POTS phone, or in another way). There may be sufficient power available from the carrier to supply NT-2 and even multiple terminals over an S interface. Terminal power from the CO

Power Supply 1

Powering TE from the NT over the Signal Wires at the S/T Interface of the BRI

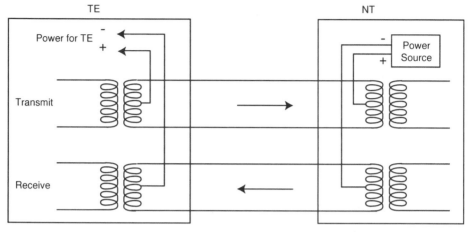

Figure 5-5

is appropriate where, for example, the TE is a telephone that should operate during a loss of local commercial power. In the US, NT-1 should be powered from a local commercial source.

Terminal Power at S/T

The first choice for S/T power on a point-topoint connection (Power Supply-1, PS-1) is to feed power from NT to TE over a phantom loop made from the two pairs that carry the transmit and receive signals. Center taps on the transformers that terminate both ends of the signal pairs provide current feeds and drains *(Fig. 5-5)*. PS-1 uses four wires, so it has half the resistance of the PS-2 and PS-3 circuits based on single conductors in the S/T cable.

Alternatively, the NT may provide power supply 2 (PS-2) on pins 7 (-) and 8 (+) or take power on pins 1 (+) and 2 (-). The TE is the opposite: it takes power on 7 and 8, or gives power on 1 and 2.

In NI-1, a reversal of battery polarity is the recommended signal that first-choice power is lost and back up or restricted power is supporting operations. The polarity reversal is to be taken as a sign by equipment that it should power-down to standby status when it can. This powering action should not affect call handling as long as sufficient battery charge remains. Terminals are to retain their internal parameter settings, Service Profile Identification (SPID), directory number (DN), etc.

Powering NT-1s from the U Interface

US regulations require the customer to provide NT-1. Bellcore recommended that the customer also supply the power for it, which no longer differs in this regard from the ITU and ANSI standard (Appendix H of ANSI's T1.601). The preferred way to power a group of co-located NTs is from the U interface, using local commercial power. Current is supplied to pins 7 (-) and 8 (+) at 48 V DC (positive ground unless floating). Bellcore recommends a power bus on the punchdown block that terminates local loops (U interfaces, *Fig. 5-6)* and converts those 2-wire lines to 8-wire modular connectors on which pins 7 and 8 are connected to power. Standard 8-wire cords then connect the line and power to NT-1s.

Any device wired to the punchdown block may draw power from it. This block then offers a single point where the customer can supply backup power as well. It is strongly recommended that a -48 V battery backup be provided, especially if there are many NT-1s in a single location. If there is a backup battery, leads 1 and 2 may be used to tell one of the NTs about the condition of the charge or availability of commercial and battery power. A good battery is denoted by a voltage of <2 V across the leads (pin 1 positive); bad, missing, or unknown quality of backup power is an open (>100 kOhm) between the pins.

One NT-1 is designated to monitor the battery backup system. If the signal on pins 1 and 2 changes (to open) that NT-1 can send an alarm to the network in the overhead portion of the BRI superframe.

With all that said, and many standards written, it was likely that the NT or terminal adapter (TA) bought on the open market in 1995 very likely had an independent wall-mount transformer—still permitted as an option. It may take some time for vendors to change from this common practice they have applied to modems and CSUs for many years—does "Never" work for you?

Premises Wiring: Power Distribution

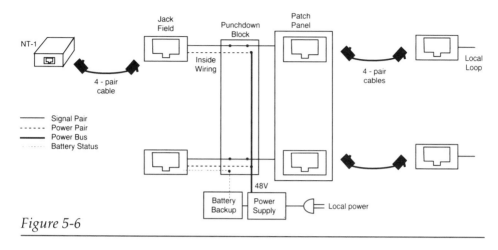

Figure 5-6

Appendix

Networking Acronyms

This glossary defines acronyms related in some way to ISDN. There are many more in telecom that are common enough for a network manager to come across, but the list has grown beyond an appendix.

Among the true acronyms listed here in all capitals are some abbreviations which normally appear in lower case letters. Numerical items are listed last.

Note that the Index doesn't cover this glossary.

Many items carry a source reference in parentheses. (802.x) = IEEE LAN standard; (Tel) = telephone terminology; (SS7) = signaling system 7; (Layer x) = in OSI protocol model; (A.NNN) = CCITT recommendation.

A

AAP   Alternate Access Provider, carrier other than local telco that can provide local loop into IXC or LEC.

AAR   Automatic Alternate Routing, failure recovery.

ABAM   Order code for 22 AWG shielded twisted pair cable used for manual cross connects of DS-1 through DS-2.

A,B,C,D   Signaling bits, for robbed bit signaling with ESF; only A and B are available with SF (Tel).

ABM   Asynchronous Balanced Mode (Layer 2).

ABS   Alternate Billing Service; credit card, 3rd party, etc. (SS7).

a.c.   Alternating Current, the form of analog voice signals, ringing, and power lines.

ACCTS   Aviation Coordinating Committee for Telecommunications Services, part of ARINC.

ACD   Automatic Call Distributor, PBX function or machine to spread calls among phones.

ACF   Access Coordination Function, tariffed service where AT&T obtains local loops between customer premises and AT&T serving office.

ACF   Advanced Communications Function, SNA software.

ACK   Positive Acknowledgment, message or control bytes in a protocol.

ACL   Applications Connectivity Link, Siemens' PHI.

ACM   Address Complete Message, signaling packet equivalent to ring-back tone or answer (SS7).

ACSE   Association Control Service Element (OSI).

ACT   ACTivate, BRI control bit.

ACT   Applied Computer Telephony, Hewlett-Packard's PHI.

A/D   Analog to Digital, usually a conversion of voice to digital format; also ADC.

ADCCP   Advanced Data Communications Control Procedure, ANSI counterpart to HDLC.

ADCR   Alternate Destination Call Redirection, service diverts calls to second site (AT&T).

ADCU   Association of Data Communications Users.

ADM   Add/Drop Multiplexer, node with 2 aggregates that supports data pass-through.

ADM   Asynchronous Disconnected Mode (Layer 2).

ADNS   ARINC Data Network Service, a packet network.

ADPCM   Adaptive Differential Pulse Code Modulation, a form of voice compression that typically uses 32 kbit/s.

AFI Authority and Format Identifier, part of network level address header (MAP).

AFIPS American Federation of Information Processing Societies.

AFR Alternate Facility Restriction.

AIB Alarm Indication Bit, BRI control bit.

AIN Advanced Intelligent Network, carrier offering more than 'pipes' to users.

AIS Alarm Indication Signal, unframed all 1's (Blue Alarm) sent downstream (to user) from fault site. Also used as T-1 keep-alive signal.

AMI Alternate Mark Inversion, line coding for T-1 spans where 0 (space) is no voltage and successive 1s (marks) are pulses of opposite polarity. See also DMC, NRZ, 4B/5B.

AMIS Audio Messaging Interchange Specification, for voice mail.

AMPS Advanced Mobile Phone Services (System), analog cellular in N.A.

AN Access Node.

ANI Automatic Number Indication, display of calling number on called phone.

ANM ANswer Message, signaling packet returned to caller indicating called party is connected (SS7).

ANR Automatic Network Routing (APPN).

ANSI American National Standards Institute, the US member of the ISO.

ANT Alternate Number Translation, ability to reroute 1-800 calls on NCP failure.

AOS Alternate Operator Service, non-telco firm responding to "dial zero."

AP Action point, SDN switch located closest to customer site.

APDU Application PDU (OSI).

API Application Programming Interface, software module or commands to separate OS or network from application.

APPC Advanced Program to Program (Peer-to-Peer) Communications, session level programming interface (APPN).

APPN Advanced Peer-to-Peer Networking, IBM networking architecture.

APS Automatic Protection Switch.

AR Access Rate, speed of a channel into a backbone network.

ARAP Apple Remote Access Protocol.

ARD Automatic Ring Down, lifting your phone rings the far end without dialing.

ARINC Aeronautical Radio, Inc., operator of private airline networks.

ARP Address Resolution Protocol, a way for routers to adjust addresses between different protocols or domains.

ARPA Advanced Research Projects Agency, created Arpanet packet network, folded into NSFnet in 1990.

ARQ Automatic Repeat reQuest, for retransmission; an error correction scheme for data links, used with a CRC.

ASAI Adjunct Switch Applications Interface, AT&T's PHI.

ASCII American Standard Code for Information Interchange, based on 7 bits plus parity.

ASDS Accunet Spectrum of Digital Services, AT&T's fractional T-1.

ASE Applications Service Element, protocol at upper layer 7 (SS7, OSI).

ASIC Application Specific Integrated Circuit.

ASN.1 Abstract Syntax Notation #1, language for managing network elements.

ASR Automatic Send/ Receive, a printer with keyboard or a Teletype machine.

ASTN Alternate Signaling Transfer Network, a CCS6 that backs up CCS7.

ATN Aeronautical Telecommunication Network, as in ATN Protocol Architecture used by ARINC.

AU Administrative Unit, payload plus pointers (SDH).

AUG AU Group, one or more AUs to fill an STM (SDH).

AUI Attachment Unit (Universal) Interface, standard connector between MAU and PLS (802.x).

Autovon Automatic Voice network, a U.S. military net.

AVS Available Seconds, when BER of line has been less than 10^{-3} for 10 consecutive seconds until UAS start.

AWG American Wire Gauge, conventional designator of wire size.

B

B Bearer channel, a DS-0 for user traffic.

B1 SOH byte carrying BIP-8 parity check (SONET).

B2 LOH byte carrying BIP parity check.

B3 POH byte carrying BIP parity check.

B3ZS Binary 3-Zero Substitution, line coding for DS-3 signal.

B8ZS Binary 8-Zero Suppression, substitutes 000+-0-+ for 00000000 to maintain ones density on T-1 line.

BATT Battery, the -48 (-40 to -52) V d.c. supply in the CO.

Bc Committed Burst, amount of data allowed in time T=Bc/CIR without being marked DE.

BCC   Block Check Code, a CRC or similarly calculated number to find transmission errors.

BCD   Binary Coded Decimal, 4-bit expression for 0 (0000) to 9 (1001).

BCM   Bit Compression Mux, same as M44 for ADPCM.

BCN   Beacon, frames sent downstream by station on ring when upstream input is lost (802.5).

B-DCS   Broadband Digital Cross-connect System, DACS OC-1, STS-1, DS-3 and higher rates only (see W-DCS).

BDLC   Burroughs Data Link Control, layer 2 in Burroughs Network Architecture.

Be   Excess Burst, transient capacity above CIR and Bc in FR net.

BECN   Backward Explicit Congestion Notification, signaling bit in frame relay header.

BER   Bit Error Ratio (Rate), errored bits over total bits; should be < 10^{-7} for transmission lines.

BERT   Bit Error Rate Test(er).

BES   Bursty Errored Second, from 1 to 319 CRC errors in ESF framing, ESB.

BEtag   Beginning/End tag; same sequence number put at head, tail of L3-PDU.

BGF   Basic Global Functions, requirements for ISDN.

BIB   Backward Indicator Bit, field in SUs (SS7).

BIP-x   Bit Interleaved Parity, error checking method where each of x bits is parity of every x^{th} bit in data block (x=8 in SONET, 16 in ATM).

B-ISDN Broadband ISDN, generally ATM access at more than 100 Mbit/s.

BISYNC Binary Synchronous communications, a protocol.

BITS   Building Integrated Timing Supply, stratum 1 clock in CO.

BIU   Basic Information Unit, up to 256 bytes of user data (RU) with RH and TH headers(SNA).

BLU   Basic Link Unit, data link level frame (SNA).

BNA   Burroughs Network Architecture, comparable to SNA.

BOC   Bell Operating Company, a telephone company.

BoD   Bandwidth on Demand, dynamic allocation of line capacity to active users.

BOM   Beginning of Message, type of segment (cell) that starts a new MAC frame, before COM and EOM (SMDS).

BOM   Bill Of Materials, list of all parts in an assembly.

BONDING Bandwidth On Demand INteroperability Group, makers of inverse muxes and standard they adopted.

BORSHT Battery feed, Over-voltage protection, Ringing, Signaling, Hybrid, Test; classic functions of analog interface.

bps Bits per second, serial digital stream data rate, now bit/s.

BPV Bipolar Violation, two pulses of the same polarity in a row.

BR Bureau of Radiocommunications, part of ITU that allocates international spectrum.

BRA Basic Rate Access, ISDN 2B+D loop.

BRI Basic Rate Interface, 2B+D on one local loop.

BSC Binary Synchronous Communications, a half-duplex protocol.

BSN Backward Sequence Number, sequence number of packet (SU) expected next (SS7).

BSP Bell System Practice, format for documents associated with equipment used by telcos.

BSRF Basic System Reference Frequency, formerly Bell SRF; Stratum 1 clock source of 8 kHz.

BSS Broadband Switching System, cell-based CO switch for B-ISDN.

BT British Telecom, primary phone company in the United Kingdom.

BTag Beginning Tag, field in header of frame whose value should match ETag.

BTAM Basic Telecommunications Access Method, older IBM mainframe comm software.

BTU Basic Transmission Unit, LU data frame of RU with RH (SNA).

BWB Bandwidth Balancing, method to reduce a station's access to a transmission bus, to improve fairness (802.6).

C

C Capacitance, the property of a device that holds an electrical charge.

C-Plane Control Plane, out of band signaling system for U-Plane.

CA*net Canadian Academic Network.

CAD Computer Aided Design, drafting on computers.

CAE Computer Aided Engineering.

CAM Computer Aided Manufacturing.

CAP Carrierless Amplitude and Phase modulation, a modem technique applied to 50 Mbit/s LAN.

CAP Competitive Access Provider, alternative to LEC for local loop to IXC or for dial tone.

CASE Common Application Service Elements, application protocol (MAP).

CAT   Category, often with a number (CAT-3) to indicate grade, as of UTP wiring.

CATV   Community Antenna Television, cable TV.

CBEMA   Computer and Business Equipment Manufacturers Association.

CBR   Constant (Continuous) Bit Rate, channel or service in ATM network for PCM voice or sync data in a steady flow with low variation in cell relay. Emulates TDM channel.

CBX   Computerized Branch eXchange, PABX.

CC   Cluster Controller, for group of dumb terminals (SNA).

CCBS   Completion of Calls to Busy Subscribers, supplementary service defined for ISDN.

cch   Connections per Circuit-hour, in Hundreds.

CCIS   Common Channel Inter-office Signaling.

CCITT   Comite Consultatif Internationale de Telegraphique et Telephonique, The International Telegraph and Telephone Consultative Committee, part of ITU that was merged with CCIR in 1993 to form TSS.

CCIR   International Radio Consultative Committee, sister group to CCITT, became part of TSS (1993).

CCIS   Common Channel Interoffice Signaling, earlier name for CCSS.

CCR   Commitment, Concurrency, and Recovery (OSI).

CCR   Customer Controlled Reconfiguration, of T-1 lines via DACS switching.

CCS   Common Channel Signaling.

CCS   Common Communication Subsystem, level 7 applications services (SNA).

CCSA   Common Control Switching Arrangement.

CCSS   CCS System, usually with a number.

CCS6   CCS system 6, first out of band signaling system in N.A. (CCIS).

CD   Carrier Detect, digital output from modem when it receives analog modem signal.

CD   Count Down, a counter that holds the number of cells queued ahead of the local message segment (802.6).

CDMA   Code Division Multiple Access, spread spectrum; broadcast frequency changes rapidly in pattern known to receiver.

CDPD   Cellular Digital Packet Data.

CELP   Code-Excited Linear Predictive coding, a voice compression algorithm used at 8 kbit/s.

CEN   Committee for European Standardization.

CENELEC Committee for European Electrotechnical Standardization.

CEPT   Conference on European Posts & Telecommunications (Conference of European Postal and Telecommunications administrations), a body that formerly set policy for services and interfaces in 26 countries.

CERN   Nuclear research facility (particle physics) in Geneva, Switzerland, where web browsing was invented.

CES   Circuit Emulation Service, ATM net pretending to be a TDM net.

CES   Connection Endpoint Suffix, number added by TE to SAPI to make address for connection; mapped to TEI by L2.

CFA   Carrier Failure Alarm, detection of red (local) or yellow (remote) alarm on T-1.

CGA   Carrier Group Alarm, trunk conditioning applied during CFA.

CHAP   Challenge Handshake Authentication Protocol, log-in security procedure for dial-in access.

CI   Congestion Indication.

CI   Connection Identifier, frame or cell address.

CIB   CRC Indication Bit, 1 if the CRC is present, 0 if it is not used (SMDS).

CICS   Customer Information Control System, IBM mainframe comm software with data base.

CIDR   Classless Inter-Domain Routing (IP).

CIR   Committed Information Rate, minimum throughput guaranteed by FR carrier.

CIT   Computer Integrated Telephony, DEC's PHI.

CL   Common Language, Bellcore codes to identify equipment, locations, etc.

CL   ConnectionLess.

CLASS   Custom(er, ized) Local Area Signaling Services, like ANI, call waiting, trace, call forwarding, etc. (SS7).

CLEI   Common Language Equipment Identifier, unique code assigned by Bellcore for label on each CO device.

CLID   Calling Line IDentification, ANI.

CLIP   Calling Line Identity Presentation, ISDN-UP service to support ANI.

CLIR   Calling Line Identity Restriction, feature where caller prevents ANI (ISDN-UP).

CLIST   Command List, similar to .BAT file (SNA).

CLLM   Consolidated Link Layer Management (802).

CLN   ConnectionLess Network, packet address is unique and network routes all traffic.

CLNAP   CLN Access Protocol.

CLNIP CLN Interface Protocol.

CLNP ConnectionLess mode Network (layer) Protocol (SONET); see CNLP.

CLNS ConnectionLess mode Network (layer) Service, ULP (SNA).

CLP Cell Loss Priority, signaling bit in ATM cell (1=low).

CLR Cell Loss Ratio (ATM).

CLS Connectionless Service.

CLTS ConnectionLess Transport Service, OSI datagram protocol.

CMA Communications Managers Association.

CMD Circuit Mode Data, ISDN call type.

CMDR Command Reject, similar to FRMR (HDLC).

CMI Coded Mark Inversion, line signal for STS-3.

CMI Constant Mark, Inverted; line coding for T-1 local loop in Japan.

CMIP Common (network) Management Information Protocol, part of the OSI network management scheme, connection oriented.

CMIS Common (network) Management Information Service, runs on CMIP (OSI).

CMISE CMIS Element.

CMOL CMIP Over LLC, reduced NMS protocol stack.

CMOS Complementary Metal Oxide Semiconductor, low power method (lower than NMOS) to make ICs.

CMOT CMIP over TCP/IP.

CMT Connection Management, part of SMT that establishes physical link between adjacent stations (FDDI).

CND Calling Number Delivery, another name for CLID, one of the CLASS services.

CNF Confirmed (OSI).

CNIS Calling Number Identification Service, provide, screen, or deliver CPN or caller ID (ISDN).

CNLP Connectionless Protocol.

CNLS Connectionless Service.

CNM Communications Network Management (SNA).

CNM Customer Network Management (Bellcore).

CO Central Office, of a phone company, where the switch is located; the other end of the local loop opposite CP.

C-O Connection Oriented.

COC Central Office Connection, separately tariffed part of circuit within a CO.

COCF Connection Oriented Convergence Function, MAC-layer entity.

CODEC COder-DECoder, converts analog voice to digital, and back.

COFA   Change of Frame Alignment, movement of SPE within STS frame.

CO-LAN Central Office Local Area Network, a data switching service based on a data PBX in a carrier's CO.

comm   Communications.

CON   Connection-Oriented Network, defines one path per logical connection.

CONP   Connection mode Network layer Protocol.

CONS   Connection-Oriented Network Services, ULP (SNA).

COS   Class Of Service.

COS   Corporation for Open Systems, R&D consortium to promote OSI in the US; see SPAG.

COSINE Cooperation for Open Systems Interconnection Networking in Europe.

COT   Central Office Terminal, equipment at CO end of multiplexed digital loop or line.

COT   Customer-Originated Trace, sends CPN to telco or police (CLASS).

CP   Central Processor, CPU that runs network under center-weighted control.

CP   Control Point, function in APPN node for routing, configuration, directory services.

CP   Customer Premises, as opposed to CO.

CPAAL   Common Part AAL, may be followed by a number to indicate type.

CPE   Customer Premises Equipment, hardware in user's office.

CPI   Computer-PBX Interface, a data interface between NTI and DEC.

CPIC   Common Programming Interface for Communications, a software tool for using LU6.2 adopted by X/Open as a standard.

CPN   Calling Party Number, DN of source of call (ISDN).

CPN   Customer Premises Node, CPE.

CPNI   Customer Proprietary Network Information, customer data held by telcos.

CPSS   Control Packet Switching System, subnetwork of supervisory channels (Newbridge).

CPU   Central Processor Unit, the computer.

CR   Carriage Return, often combined with a line feed when sending to a printer.

CRC   Cyclic Redundancy Check, an error detection scheme, for ARQ or frame/cell discard.

CRF   Connection Related Function (ISDN).

CRIS   Customer Records Information System, telco OSS.

CRT Cathode Ray Tube, simple computer terminal.

CRV Coding Rule Violation, unique bit signal for F bit in frame 1 of CMI.

CS Circuit Switched, uses TDM rather than packets.

CS Convergence Sublayer, where header and trailer are added before segmentation (ATM).

CSA Carrier Service Area, defined by a local loop length out of a CO, or from a remote switch unit or multiplexer.

CSA Callpath Services Architecture, for PBX to IBM host interface.

CSC Circuit-Switched Channel (Connection).

CSDC Circuit Switched Digital Capability, AT&T version of Sw56.

CSMA Carrier Sense Multiple Access, a LAN transport method, usually with "/CD" for collision detection or "/CA" collision avoidance; LAN protocols at physical layer.

CSO Cold Start Only, BRI control bit.

CSPDN Circuit Switched Public Data Network.

CS-PDU Convergence Layer PDU, info plus new header and trailer to make packet that is segmented into cells or SUs.

CSTA Computer Supported Telephony Application, PHI from ECMA.

CSU Channel Service Unit, the interface to the T-1 line that terminates the local loop.

CT2 Cordless Telephone, second version; digital wireless telephone service defined by ETSI.

CTR Common Technical Requirements, European standards.

CTS Clear To Send, lead on interface indicating DCE is ready to receive data.

CUG Closed User Group.

CV Coding Violation, transmission error in SONET section.

CVSD Continuously Variable Slope Delta modulation, a voice encoding technique offering variable compression.

CWC Center-Weighted Control, A central processor runs a network-wide functions while nodes do local tasks.

D

D Delta (or Data) channel, 16 kbit/s in BRI, 64 kbit/s in PRI, used for signaling (and perhaps some packet data).

D3 Third generation channel bank, 24 channels on one T-1.

D4 Fourth generation digital channel bank, up to 48 voice channels on two T-1s or one T-1C.

D5 Fifth generation channel bank with ESF.

DA	Destination Address, field in frame header (802).	**dBrn**	Power level relative to noise, dBm + 90.
D/A	Digital to Analog, decoding of voice signal from bits to sound; also DAC.	**dBrnC**	dBrn through a C-weighted audio filter (matches ear's response).
D/A	Drop and Add, similar to drop and insert.	**DBS**	Direct Broadcast Satellite.
DACS	Digital Access and Cross-connect System, a digital switching device for routing T-1 lines, and DS-0 portions of lines, among multiple T-1 ports.	**DB25**	25-pin connector specified for RS-232 I/F.
		d.c.	Direct Current, used for some signaling forms; type of power in CO.
DAMA	Demand Assigned Multiple Access, multiplexing technique to share satellite channels (Vsat).	**DCA**	Distributed Communications Architecture, networking scheme of Sperry Univac.
DARA	Dynamic Alternate Routing Algorithm.	**DCC**	Data Communications Channel, overhead connection in D bytes for SONET management.
DARPA	Defense ARPA, formerly just ARPA.	**DCC**	Digital Cross Connect, generic DACS.
DAS	Dual-Attached (Access) Station, device on main dual FO rings, 4 fibers (FDDI).	**DCC**	Direct Connect Card, data interface module on a T-1 bandwidth manager.
DASD	Direct Access Storage Device (SNA).	**DCCID**	Data Compression Context Identifier, 'session' number to support multiple symbol dictionaries.
DASS	Digital Access Signaling System, protocol for ISDN D channel in U.K.	**DCE**	Data Circuit-terminating Equipment, see next DCE.
dB	Decibel, 1/10th of a bel; $10 \log (x/y)$ where x/y is a ratio of like quantities (power).	**DCE**	Data Communications Equipment, 'gender' of interface on modem or CSU; see DTE
dBm	Power level referenced to 1 mW at 1004 Hz into 600 ohms impedance.	**DCS**	Digital Cross-connect System, DACS.
dBm0	Power that would be at zero TLP reference level, = measurement - (TLP at that point).	**DDCMP**	Digital Data Communications Message Protocol.

DDD   Direct Distance Dialing, refers to PSTN.

DDS   Digital Data System, network that supports DATAPHONE Digital Service.

DDSD   Delay Dial Start Dial, a start-stop protocol for dialing into a CO switch.

DE   Discard Eligibility, bit in FR header denoting lower priority; as when exceeding CIR or Bc.

DEA   DEActivate, BRI control bit.

DECmcc   Digital Equipment Corp. Management Control Center, umbrella network management system.

DES   Data Encryption Standard, moderately difficult to break.

DFC   Data Flow Control, layer 5 of SNA.

DFN   Deutsche Forschungsnetz Verein, German Research Network Association.

DGM   Degraded Minute, time when BER is between 10-6 and 10-3.

D/I   Drop and Insert, a mux function or type.

DID   Direct Inward Dial, CO directs call to specific extension on PBX, usually via DNIS.

DIP   Dual In-line Package, for chips and switches.

DIS   Draft International Standard, preliminary form of OSI standard.

DISA   Direct Inward System Access, PBX feature that allows outside caller to use all features, like calling out again.

DISC   Disconnect, command frame sent between LLC entities (Layer 2).

DL   Data Link.

DLC   Data Link Connection, one logical bit stream in LAPD (Layer 2).

DLC   Data Link Control, level 2 control of trunk to adjacent node (SNA).

DLC   Digital Loop Carrier, mux system to gather analog loops and carry them to CO.

DLCI   Data Link Connection Identifier, address in frame relay (I.122); LAPD address consisting of SAPI and TEI.

DLE   Data Link Escape, ESC.

DLL   Data Link Layer, layer 2 (OSI).

DLS   Data Link Switching, IBM way to carry Netbios and spoofed SDLC over TCP/IP.

DM   Disconnected Mode, LLC frame to reject a connection request (Layer 2).

DMC   Differential Manchester Code, pulse pattern that puts transition at center of each bit time for clocking, transition [none] at start of period for 0 [1] (802.5).

DME   Distributed Management Environment, OSF's network management architecture.

DMI	Digitally Multiplexed Interface, AT&T interface for 23 64 kbit/s channels and a 24th for signaling; precursor to PRI.	**DPC**	Destination signal transfer Point Code, level 3 address in SU of STP (SS7).
DMPDU	Derived MAC Protocol Data Unit, a 44-octet segment of upper layer packet plus cell header/trailer (802.6); see L2PDU.	**DPCM**	Differential Pulse Code Modulation, voice compression algorithm used in ADPCM.
DMS	Digital Multiplex System.	**DPGS**	Digital Pair Gain System, multiplexer and line driver to put 2+ channels over 1 or 2 pair of wires.
DN	Directory Number, network address used to reach called party (POTS, ISDN).	**DPNSS**	Digital Private Network Signaling System, PBX interface for common channel signaling.
DNA	Digital Network Architecture, DEC's networking scheme.	**DPO**	Dial Pulse Originate, a form of channel bank plug-in that accepts dial pulses.
DNIC	Data Network Identification Code, assigned like an area code to public data networks.	**DPT**	Dial Pulse Terminate, a channel bank plug that outputs pulses.
DNIC	Data Network Interface Circuit, 2B+D ISDN U interface.	**DQDB**	Distributed Queue Dual Bus, an IEEE 802.6 protocol to access MAN's, typically at T-1, T-3, or faster.
DNIS	Dialed Number Identification Service, where carrier delivers number of called extension after PBX acknowledges call.	**DS-0**	Digital Signal level 0, 64,000 bit/s, the worldwide standard speed for PCM digitized voice channels.
DNR	DCE Not Ready, signaling bit in CMI.	**DS-0A**	Digital Signal level 0 with a single rate adapted channel.
DoV	Data over Voice, modems combine voice and data on one twisted pair.	**DS-0B**	Digital Signal level 0 with multiple channels sub-rate multiplexed in DDS format.
DP	Draft Proposal, of an ISO standard.	**DS-1**	Digital Signal level 1, 1.544 Mbit/s in North America, 2.048 Mbit/s in CCITT countries.
DP	Dial Pulse, rotary dialing rather than DTMF.		
DPBX	Data PBX, a switch under control of end users at terminals.		

DS-1A   Proposed designation for 2.048 Mbit/s in North America.

DS-1C   Two T-1s, used mostly by Telcos internally.

DS-2   Four T-1s, little used in US, common in Japan.

DS-3   Digital Signal level 3, 44.736 Mbit/s, carrying 28 T-1s.

DSAP   Destination Service Access Point, address field in header of LLC frame to identify a user within a station address (Layer 2).

DSG   Default Slot Generator, the function in a station that marks time slots on the bus (802.6).

DSI   Digital Speech Interpolation, a voice compression technique that relies on the statistics of voice traffic over many channels.

DSL   Digital Subscriber Line (ISDN).

DSP   Digital Signal Processor, specialized chip optimized for fast computations.

DSP   Display System Protocol, protocol for faster bisync traffic over packet nets.

DSR   Data Set Ready, signal indicating DCE and line ready to receive data.

DSS1   D-channel (Digital Subscriber) Signaling System 1, access protocol for switched connection signaling from NT to ISDN switch (Q.931 & ANSI T1S1/90-214).

DSU   Digital (Data) Service Unit, converts RS-232 or other terminal interface to line coding for local loop transmission.

DSX-1   Digital Signal cross connect, level 1; part of the DS-1 specification, T-1 or E-1.

DSX-z   Digital Signal cross connect where 'z' may be 0A, 0B, 1, 1C, 2, 3, etc. to indicate the level.

DT   Data Transfer, type of TPDU in ISDN.

DTAU   Digital Test Access Unit, CO equipment on T-1 line.

DTE   Data Terminal Equipment, 'gender' of interface on terminal or PC; see DCE.

DTI   Digital Trunk Interface, T-1 port on Northern Telecom PBX.

DTMF   Dual Tone Multi-Frequency, TOUCHTONE dialing, as opposed to DP.

DTP   Data Transfer Protocol.

DTR   Data Terminal Ready, signal that terminal is ready to receive data from DCE.

DTU   Data Termination Unit, Newbridge TA for data.

DVI   Digital Video Interactive, applications with large, bursty bandwidth.

DX   Duplex, a 2-way audio channel bank plug without signaling.

DXI   Data Exchange Interface, serial protocol for SNMP for any speed.

E

E-1 European digital signal level 1, 2.048 Mbit/s.

E-ADPCM Embedded ADPCM, packetized voice with "core" and "enhancement" portion to each frame (G.726).

EA Extended Address, or address extension bit, =1 in last byte of frame header.

EASInet Network for European Academic Supercomputer Initiative.

E&M Signaling leads on a voice tie line, known as Ear and Mouth.

EBCDIC Extended Binary Coded Decimal Interchange Code, extended character set on IBM hosts.

EC Error Correction, process to check packets for errors and send again if needed.

EC Enterprise Controller, new terminal cluster controller, e.g. 3174.

EC European Community, covered by common telecom standards.

ECC Error Checking Code, 2 bytes (usually) in frame or packet derived from data to let receiver test for transmission errors.

ECHO European Clearing House Organization, foreign exchange settlement network, run by SWIFT.

ECL Emitter Coupled Logic, transistor circuit type optimized for high speed.

ECMA European Computer Manufacturers Association.

ECN Explicit Congestion Notification, network warns terminals of congestion by setting bits in frame header (I.122).

ECO Engineering Change Order, document from designer ordering change in product.

ECTRA European Committee on (Telecom) Regulatory Affairs, created out of CEPT in 1990 to be regulatory half, as opposed to operational part, of carriers and PTTs.

ECSA Exchange Carrier Standards Association.

ED Ending Delimiter, unique symbol to mark end of LAN frame (TT in FDDI, HDLC flag, etc.).

EDI Electronic Document (Data) Interchange, transfer of business information (P.O., invoice, etc.) in defined formats.

EDSX Electronic DSX, usually followed by a "-N" for signal level.

EETDN End to End Transit Delay Negotiation, part of call setup via X.25 (ISDN).

EFT Electronic Funds Transfer.

EGP   Exterior Gateway Protocol, on the Internet (TCP/IP).

EIA   Electronic Industries Association, publisher of standards (e.g. RS-232).

EISA   Extended ISA, 32-bit PC bus compatible with AT style PCs.

EKTS   Electronic Key Telephone Service, ISDN terminal mode.

EMA   Enterprise Management Architecture, DEC's umbrella network management system.

EMI   Electro Magnetic Interference.

EMS   Element Management System, usually a vendor-specific NMS for a hardware domain (OSI).

EN   End Node, limited capability access device (APPN).

ENQ   Enquiry, control byte that requests a repeat transmission or control of line.

ENTELEC   Energy Telecommunications and Electrical Association.

EOC   Embedded Operation Channel, D bytes devoted to alarms, supervision, and provisioning (SONET); control field in M channel of BRI.

EOM   End of Message, cell type carrying last segment of frame.

EOT   End Of Transmission, control byte; preceded by DLE indicates switched station going on hook.

EPSCS   Enhanced Private Switched Communications Service.

ERL   Echo Return Loss.

ERLE   ERL Enhancement, reduction in echo level produced by echo canceller.

ERS   Errored Second, a 1 sec. interval containing 1 or more transmission errors.

ESB   Errored Second type B, new name for bursty ES.

ESC   Escape, an ASCII character.

ESD   ElectroStatic Discharge, electrical "shock" from person or other source that can destroy semi-conductors.

ESF   Extended Super Frame, formerly called F_e.

ESP   Enhanced Service Provider, a firm that delivers its product over the phone.

ESS   Electronic Switching System, a CO switch.

ET   Exchange Termination, standards talk for ISDN switching equipment in CO.

ETag   End Tag, field in trailer of frame whose value should match that in BTag.

ETC   End of Transmission Block, control byte in BSC.

ETN   Electronic Tandem Network.

ETO Equalized Transmit Only, voice interface with compensation to correct for frequency response of the line.

ETS Electronic Tandem Switching.

ETSI European Telecommunications Standards Institute, coordinates telecommunications policies.

ETX End of Text, control byte.

F

F Final, control bit in frame header (Layer 2).

F Framing, bit position in TDM frame where known pattern repeats.

F1 Flow (of OA&M cells) at level 1, over a SONET section (ATM).

F2 Flow of OA&M cells over a line.

F3 Flow of OA&M cells between PTEs.

F4 Flow of OA&M cells for metasignaling and VP management.

F5 Flow of OA&M cells specific to a logical connection on one VPI/VCI.

FACS Facility Assignment Control System, for telco to manage outside plant (local loops).

FAD Factory Authorized Dealer.

FADU File Access Data Unit (OSI).

FAS Facility Associated Signaling, D channel is on same interface as controlled B channels (ISDN).

FAS Frame Alignment Signal, bit or byte used by receiver to locate TDM channels.

FASTAR Fast Automatic Restoration, of DS-3s via DACS switching (AT&T).

FAX Facsimile.

FB Framing Bit.

FC Frame Control, field to define type of frame (FDDI).

FCC Federal Communication Commission, regulates communications in US.

FCOT Fiber optic Central Office Terminal.

FCS Frame Check Sequence, error checking code like CRC (Layer 2).

FDDI Fiber Distributed Data Interface, 100 Mbit/s FO standard for a LAN or MAN.

FDL Facility Data Link, part of the ESF framing bits available for user data, in some cases.

FDM Frequency Division Multiplexer.

FDMA Frequency Division Multiple Access, wireless access where each channel has separate radio carrier frequency.

FDX Full DupleX, simultaneous transmission in both directions.

F$_e$ Extended framing ("F sub e"), old name for ESF.

FEA   Far End Alarm, repeating bit C-3 in DS-3 format identifies alarm or status.

FEBE   Far End Block Error, alarm signal (count of BIP errors received; ATM uses Z2 byte in SONET LOH).

FEC   Forward Error Correction, using redundancy in a signal to allow the receiver to correct transmission errors.

FECN   Forward Explicit Congestion Notification, signaling bit in frame relay header.

FEP   Front End Processor, peripheral computer to mainframe CPU, handles communications.

FERF   Far End Receive Failure, alarm signal (ATM).

FEXT   Far End Cross Talk.

FGA   Feature Group A, set of signaling and other functions at LEC-IXC interface; also defined for FGB to FGD.

FH   Frame Handler, term in standards for FRS or network.

FIB   Forward Indicator Bit, field in SUs (SS7).

FID   Format Identification, bit C-1 in DS-3 format shows if M13, M28, or Syntran signal.

FIFO   First In First Out, buffer type that delays bit stream but preserves order.

FIPS   Federal Information Processing Standards, for networks.

FISU   Fill-In Signaling Unit, 'idle' packet that carries ACKs as sequence numbers (SS7).

FITL   Fiber In The Loop, optical technology from CO to customer premises.

FIX   Federal Internet Exchange, point of interconnection for U.S. agency research networks.

FMBS   Frame Mode Bearer Service, FR on ISDN.

FO   Fiber Optic, based on optical cable.

FOTP   Fiber Optic Test Procedure.

FOTS   Fiber Optic Terminal System, mux or CO switch interface.

FPDU   FTAM PDU.

FPDU   Frame relay Protocol Data Unit (I.122).

FR   Frame Relay, interface to simplified packetized switching network (I.122, T1.617).

FRAD   Frame Relay Assembler/ Disassembler, functions like a PAD, but for frames.

FRBS   FR Bearer Service, newer name for FMBS.

FRS   Frame Relay Switch or Service.

FRSE   FR Service Emulation, as by IWF over ATM.

FRMR Frame Reject, LLC response to error that cannot be corrected by ARQ, may cause reset or disconnect (Layer 2).

FS Failed Second, now called UAS.

FSK Frequency Shift Keying, modem method where carrier shifts between two fixed frequencies in voice range.

FSN Forward Sequence Number, sent sequence number of this SU/ packet (SS7).

FT-1 Fractional T-1, digital capacity of N x 64 kbit/s but usually less than 1/2 a T-1.

FTAM File Transfer, Access, and Management; an OSI layer-7 protocol for LAN interworking (802).

FTP File Transfer Protocol (TCP/IP).

FTTC Fiber To The Curb, local loop is fiber from CO to just outside CP, wire into CP.

FUNI Frame-based UNI, format for low speed access to ATM networks.

FX Foreign Exchange, not the nearest CO. FX line goes from a CO or PBX to beyond its normal service area.

FXO Foreign Exchange, Office; an interface at the end of a private line connected to a switch.

FXS Foreign Exchange, Subscriber (or station); an interface at the end of an FX line connected to a telephone, etc.

G

G3 Group 3, analog facsimile standard at up to 9.6 kbit/s.

G4 Group 4, digital facsimile standard at 56/64 kbit/s.

GA Group Address, for multicasting.

Gbit/s Giga bits per second, billions (109) per second.

GFC Generic Flow Control, first half-byte in cell header at UNI.

GOSIP Government OSI Profile, suite of protocols mandated for US Federal and U.K. contractors; –T = Transport model; –A = Application model.

GPS Global Positioning System, satellites that report exact time.

GS Ground start, analog phone interface.

GSM Group Speciale Mobile (Global System for Mobile communications), CEPT standard on digital cellular.

GSTM General Switched Telephone Network, CCITT term to replace PSTN after 1990's privatizations.

GUI Graphical User Interface.

H

Hx High-speed bearer channels (ISDN):

H_0 384 kbit/s.

H_1 Payload in a DS-1 channel.

H_{11} 1.536 Mbit/s (N. Amer.).

H_{12} 1.920 Mbit/s (CEPT areas).

H_2 Payload of a DS-3 channel.

H_{21} 32.768 Mbit/s (CEPT).

H_{22} 43.008 Mbit/s, the payload of 28 T-1s, to 41.160 Mbit/s, if including 18 more DS-0s from the DS-3 overhead (N. A.)

H_3 Would have been 60-70 Mbit/s, but left undefined for lack of interest.

H_4 135.168 Mbit/s (88 T-1s)

HCDS   High Capacity Digital Service, Bellcore T-1 specification.

HCI   Host Command Interface, Mitel's PHI.

HCM   High Capacity Multiplexing, 6 channels of 9600 in a DS-0.

HCV   High Capacity Voice, 8 or 16 kbit/s scheme.

HDB3   High Density Bipolar 3-zeros, line coding for 2 Mbit/s lines replaces 4 zeros with BPV (CEPT).

HDLC   High-level Data Link Control, layer-2 full-duplex protocol.

HDSL   High bit-rate Digital Subscriber Line, Bellcore Standard for way to carry DS-1 over local loops without repeaters.

HDT   Host Digital Terminal, CO end of multiplexed local loop (see RDT).

HDTV   High Definition Television, double resolution TV image and candidate application for broadband networks.

HE   Header Extension, a 12-octet field for various information elements (SMDS).

HEC   Header Error Control, ECC in ATM cell for header, not data. (See HCS)

HEL   Header Extension Length, the number of 32-bit words in HE (802.6).

HEPNET   High Energy Physics Network, international R&D net.

HIPPI   HIgh-speed Peripheral Parallel Interface, computer channel simplex interface clocked at 25 MHz; 800 Mbit/s when 32 bits wide, 1.6 Gbit/s when 64 bits.

HOB   Head Of Bus, station and function that generates cells or slots on a bus (DQDB).

HPR   High Performance Routing, a form of dynamic call routing in the PSTN; IBM method for LAN routing.

HRC   Hybrid Ring Control, TDM sublayer at bottom of data link (2) that splits FDDI into packet- and circuit-switched parts.

HSSI   High Speed Serial Interface, of 600 or 1200 Mbit/s.

Hz   Hertz, cycles/second.

I

I   I class central office switch is not in HPR network.

I   Idle, line state symbol (FDDI).

I   Information, type of layer 2 frame that carries user data.

IA   Implementation Agreement, based on a subset of CCITT or ISO standards without options to ensure interoperability.

Ia   Interface point a, the network-side port of TE-1, NT-1, or NT-2 (ISDN PRI)

Ib   Interface point b, the user-side port of TE-1, NT-1, or NT-2 (ISDN PRI).

IA5   International Alphabet #5, coding for signaling information (ISDN).

IAB   Internet Activities Board, defines LAN standards like SNMP.

IACS   Integrated Access and Cross-connect System, AT&T box with DACS and mux functions via packet switching fabric.

IAD   Integrated Access Device, CPE that supports multiple services on public/private net.

IAM   Initial Address Message, call request packet (SS7).

IBR   Intermediate Bit Rate, between 64 and 1536 kbit/s; fractional T-1 rates.

IC   Integrated Circuit.

ICA   International Communications Association, a users group.

ICCF   Industry Carriers Compatibility Forum.

ICIP   InterCarrier Interface Protocol, connection between two public networks.

ICMP   Internet Control Message Protocol, reports to a host errors detected in a router by IP.

IDF   Intermediate Distribution Frame.

IDLC   Integrated Digital Loop Carrier, combination of RDT (remote mux), transmission facility, and IDT to feed voice and data into a CO switch.

IDT   Integrated Digital Terminal, M24 function in a CO switch to terminate a T-1 line from RDT.

IE   Information Element, part of a message; e.g. status of one PVC in a report.

IEC   Inter-Exchange Carrier, a long distance company, carries traffic between LATA's.

IEC   International Electrotechnical Commission, standards body.

IEEE Institute of Electrical and Electronics Engineers, Inc.; engineering society; one of the groups which set standards for communications.

IETF Internet Engineering Task Force, adopts RFCs.

I/F Interface (also i/f).

IG ISDN Gateway (AT&T).

IGOSS Industry/Government Open Systems Specification, broader GOSIP.

IGRP Interior Gateway Routing Protocol, learns best routes through LAN internet (TCP/IP).

ILS Idle Line State, presence of idle codes on optical fiber line (FDDI).

IMD InterModulation Distortion.

IMPDU Initial MAC PDU, the SDU received from LLC with additional header/ trailer to aid in segmentation and reassembly (802.6).

IN Intelligent Network.

INA Integrated Network Access, multiple services over one local loop.

INE Intelligent Network Element,

I/O Input/Output.

IOC Inter-Office Channel, portion of T-1 or other line between COs of the IXC.

IOC Isdn Ordering Code, 1 or 2-letter code for a complete set of configuration parameters for ISDN BRI service.

IMPDU Initial MAC Protocol Data Unit,

IP Internet Protocol, datagram network layer (2); basis for TCP, UDP, etc.

IPX Internetwork Packet eXchange, Novell's networking protocol, based on XNS.

IR InfraRed, light with wave length longer than red, like 1300 nm used over fiber.

IRQ Interrupt ReQuest.

IS International Standard.

ISA Industry Standard Architecture, the personal computer design based on IBM's AT model.

ISDN Integrated Services Digital Network.

ISDN-UP ISDN User Part, protocol from layer 3 and up for signaling services for users, Q.761-Q.766 (SS7).

ISDU Isochronous Service Data Unit, upper layer packet from TDM or circuit-switched service (802.6).

ISI Inter-Symbol Interference, source of errors where pulses (symbols) spread and overlap due to dispersion.

ISO International Standards Organization, ANSI is US member.

ISP Information Service Provider; later specialized to Internet Service Provider.

ISR   Intermediate Session Routing, performs address swapping and flow control (APPN).

ISSI   Inter-Switching System Interface, between nodes in a public network, not available to CPE (e.g. SMDS to B-ISDN).

ISSIP   ISSI Protocol.

ISUP   ISDN User Part (SS7).

ITB   End of Intermediate Transmission Block, control byte in BSC.

ITG   Integrated Telemarketing Gateway.

ITU   International Telecommunications Union, UN agency, parent of ITU-TS (formerly CCITT), CCIR, etc.

ITUA   Independent T1 Users Association (dissolved, 1995).

ITU-RS   ITU-Radiocommunications Sector.

ITU-T   Short for ITU-TSS

ITU-TSS   ITU Telecommunications Standardization Sector, successor to CCITT.

IVR   Interactive Voice Response.

IWF   InterWorking Function, the conversation process between FR and X.25, FR and ATM, etc.

IWU   InterWorking Unit, protocol converter between packet formats like FR and ATM.

IXC   IntereXchange Carrier, a long distance phone company or IEC, as opposed to LEC.

J

J   Non-data character for starting delimiter (11000) in 4B/5B coding (802.6).

JB7   Jam Bit 7, force bits in position 7 within a DS-0 to 1 for 1's density.

JPEG   Joint Photographic Experts Group, part of ISO that defined digital storage format for still photos (cf, MPEG).

JTC1   Joint Technical Committee 1, of IEC and ISO.

K

K   Non-data character for starting delimiter (10001) in 4B/5B coding (802.6).

k   Kilo, prefix for 1000; 1000 bit/s; K = 1024 bytes when applied to RAM size.

K2   LOH byte (SONET).

kbit/s   Thousands of bits per second

KDD   Kokusai Denshin Denwa, Japanese international long distance carrier.

KG   Key Generator (Krypto Gear), encryption equipment from NSA.

kHz   Kilohertz, thousands of cycles per second

L

L2-PDU   Layer 2 Protocol Data Unit, fixed length cell.

L3-PDU Layer 3 Protocol Data Unit, a variable length packet at OSI level 3.

LADT Local Area Data Transport, telco circuit on copper pair.

LAN Local Area Network.

LANE LAN Emulation (ATM).

LAP Link Access Procedure (Protocol), layer 2 protocol for error correcting between 2 devices.

LAPB LAP Balanced, HDLC protocol for data sent into X.25 network, etc.

LAPD Variant of LAPB for ISDN D channels.

LAPD+ LAPD protocol for other than D channels, e.g. B channels.

LAPM LAP Modem, part of V.42 modem standard.

LAT Local Area Transport, DECnet protocol for terminals.

LATA Local Access and Transport Area, a geographic region. The LEC can carry all traffic within a LATA, but nothing between LATA's.

LBO Line Build Out, insertion of loss in a short transmission line to make it act like a longer line.

LC Local Channel, the local loop.

LCD Liquid Crystal Display.

LCI Logical Connection Identifier, short address in connection-oriented frame.

LCN Logical Channel Number, form of PVC address in an X.25 packet.

LCP Link Control Protocol, part of PPP.

LD-CELP Low Delay CELP, voice compression with small processing delay (G.728).

LDC LATA Distribution Channel, line between local CO and POP.

LDM Limited Distance Modem.

LEC Local Exchange Carrier, a telco.

LED Light Emitting Diode, semiconductor used as light source in FO transmitters.

LEN Low Entry Networking, most basic subset of APPN.

LEN Local Exchange Node, CO switch of LEC.

LEO Low Earth Orbiting satellite, option for global 'cellular' phones.

LGE Loop- or Ground-start, Exchange; FXO analog voice interface.

LGS Loop- or Ground-start, Subscriber; FXS analog voice interface.

LIDB Line Identification Data Base (SS7).

LIV Link Integrity Verification (FR).

LLB Local Loop Back.

LLC Logical Link Control, the upper sublayer of the OSI data link layer (layer 2).

LLC1   Connection oriented LLC.

LLC2   Connectionless LLC.

LM   Layer Management, control function for protocol.

LME   Layer Management Entity, the process that controls configuration, etc. (802.6).

LMI   Local Management Interface, transport specification for frame relay that sets way to report status of DLCIs.

LOF   Loss Of Frame, condition where mux cannot find framing, OOF, for 2.5 sec.

LOFC   Loss of Frame Count, number of LOFs.

LOP   Loss of Pointer, SONET error condition.

LOS   Loss Of Signal, incoming signal not present (no received data).

LPC   Linear Predictive Coding, voice encoding technique.

LPDA   Link Problem Determination Aid, part of Netview NMS (SNA).

LS   Loop Start, analog phone interface.

LSAP   Link layer Service Access Point, logical address of boundary between layer 3 and LLC sublayer in 2 (802).

LSB   Least Significant Bit, position in byte with smallest value.

LSSU   Link Status Signaling Unit, control packet at layer 3 (SS7).

LT   Loop Termination, in the CO on a BRI.

LTE   Line Terminating Equipment, SONET nodes that switch, etc. and so create or take apart an SPE (SONET).

LU   Logical Unit, upper level protocol in SNA.

LU6.2   Set of services that support program to program communications.

M

M   Maintenance, overhead bits in frames and superframes at BRI.

M   Million when used as prefix to abbreviation: Mbit/s.

m   Milli (1/1000) when used as prefix: mm = millimeter

m   Meter (39.37 inches).

M13   Multiplexer between DS-1 and DS-3 levels.

M24   Multiplexer function between 24 DS-0 channels and a T-1, a channel bank.

M28   Same as M13, but different format, not compatible.

M44   Multiplexer function to put 44 ADPCM channels into one T-1; four bundles, each of one common signaling channel with 11 voice channels; transcoder or BCM.

M48   Multiplexer function to put 48 ADPCM channels into one T-1; signaling in each voice channel.

M55   ADPCM multiplexer that puts 55 voice channels in five bundles on an E-1.

MAC   Medium (Media) Access Control, the lower sublayer of the OSI data link layer.

MAN   Metropolitan Area Network, typically 100 Mbit/s.

MAP   Manufacturing Automation Protocol, for LAN's; closely related to TOP, and written MAP/TOP (802.4).

MAU   Media Access Unit, device attached physically to Ethernet cable (802.3).

MAU   Multiple (Multistation, Media) Access Unit, hub device in a TR LAN (802.5).

Mbit/s   Megabit (1,000,000 bits) per second. .

MCC   Master Control Center, part of DEC's umbrella network management system, EMA.

MCF   MAC Convergence Function, how an SDU is framed into a packet (PDU), segmented, and loaded into cells (802.6).

MCP   MAC Convergence Protocol, segmentation and reassembly procedure to put MSDUs into cells (802.6).

MDDB   Multi-Drop Data Bridging, digital bridging of PCM encoded modem signals, equivalent to analog bridging.

MDF   Main Distribution Frame, large CO wire rack for low speed data and voice cross connects.

MDI   Medium Dependent Interface, link between MAU and cable (802 Layer 1).

MF   Multi-Frequency, tone signaling on analog circuits.

MFA   MultiFrame Alignment, code in time slot 16 of E-1 to mark start of superframe.

MFJ   Modified Final Judgment, court decision that split AT&T in 1984.

MHS   Message Handling System, OSI store and forward protocol.

MHz   Megahertz, million cycles per second.

MIB   Management Information Base, OSI defined description of a network for management purposes (SNMP, IP).

MIC   Media Interface Connector, dual-fiber equipment socket and cable plug (FDDI).

MIS   Management Information Systems, dept. that runs the big computers.

MJU   Multipoint Junction Unit, a digital data bridge for DDS (DS-0B or 56 kbit/s), often part of a DACS.

MLHG   MultiLine Hunt Group, operation mode for ISDN terminal.

MLPPP   MultiLink PPP, protocol to split data stream over multiple channels.

MMFS   Manufacturing Messaging Format Standard, application protocol (MAP).

MML   Man-Machine Language, commands and responses understandable by both human and device being controlled.

MNP   Microcom Networking Protocol, error correcting protocol and compression in modems.

modem   MOdulate/DEModulate, modulate analog signal from digital data and reverse.

MPA   Manufacturing Program Analysis, evaluation standard for vendor plants (TR411).

MPEG   Motion Picture Experts Group, part of ISO that defined digital video compression and file format (cf, JPEG).

MPDU   MAC PDU (802.6).

MPL   Maximum Packet Lifetime, number of hops allowed before packet is discarded.

MPMC   Multi-Peer Multicast, "N-way" mutual broadcasting of information.

MPOA   MultiProtocol Over ATM, encapsulation formats.

MS   Management Service (SNA).

ms   Millisecond, 1/1000 second.

M/S   Master/Slave, relationship in a protocol where master always issues commands and slave only responds.

MSAP   MAC Service Access Point, logical address (up to 60 bits) of boundary between MAC and LLC sublayers (802).

MSDU   MAC Service Data Unit, data packet in LAN format; may be long and variable length before segmentation into cells.

MSS   MAN Switching System.

MSU   Message Signaling Unit, layer 3 packet (SS7).

MTA   Metallic Test Access, service point on equipment in CO.

MTBF   Mean Time Between Failures, average for one device.

MTBSO   Mean Time Between Service Outages.

MTP   Message Transfer Part, set of connectionless protocols at lower layer 3 and below (SS7); cf ISUP.

MTS   Message Toll Service, normal dial up phone service.

MTTR   Mean Time To Repair.

MTU   Maximum Transmission Unit, largest PDU.

MUX   Multiplexer.

N

N   N class central office has tandem switch that participates in HPR.

N	Digit from 2 to 9 inclusive.	**NDF**	New Data Flag, inversion of some pointer bits to indicate change in SPE position in STS frame (SONET).
n	Nano, prefix meaning 10^{-9} of the unit as nm = 10^{-9} meter.		
N.A.	North America.	**NDIS**	Network Driver Interface Specification.
NAK	Negative Acknowledgment, protocol control byte or frame indicating error (also NACK).	**NE**	Network Element, device or all similar devices in a network.
NANP	North American Numbering Plan, how area codes are assigned.	**NEBS**	Network Equipment-Building System, Bellcore generic spec for CO equipment (TR63).
NAS	Network Applications Solutions, set of DEC APIs for communication.	**NECA**	National Exchange Carriers Association.
NAU	Network Addressable Unit, addressable device or process running an SNA protocol.	**NET3**	EC standard for BRI.
		NET5	EC standard for PRI.
		NET33	EC standard for ISDN telephones.
NBS	National Bureau of Standards, now NIST.	**NEXT**	Near End Cross Talk, interference on 2-wire interfaces from sent signals leaking back into the receiver.
NCB	Network Control Block, command packet in SNMP.		
NCB	Network Control Block, transport protocol in LAN Manager (level 4).	**NFS**	Network File System, protocol for file transfers on a LAN.
NCC	Network Control Center.	**NFS**	Network File Server, computer with shared storage, on a LAN.
NCI	Network Control Interface.		
NCP	Network Control Point, for SDN and AT&T switched network.	**NHRP**	Next Hop Resolution Protocol, for routing over large clouds (IETF).
NCP	Network Control Program, software for FEP in SNA.	**NI**	Network Interface; demarcation point between PSTN and CPE.
NCP	Network Control Protocol, part of PPP.	**NIC**	Network Interface Card, add-in card for PC, etc. to connect to LAN.
NCTE	Network Channel Terminating Equipment; first device at CP end of local loop; e.g., CSU.	**NID**	Network IDentification, field in network level header (MAP).

NISDN	Narrowband ISDN, access at T-1 or less.	**NPDA**	Network Problem Determination Application, fault isolation software for IBM hosts, part of NetView.
NIST	National Institute of Standards and Technology, name change for National Bureau of Standards.	**NPDU**	Network PDU, layer 3 packet (OSI).
NIUF	North-american Isdn Users Forum, a group associated with NIST.	**NPI**	Numbering Plan Indicator, field in message with DN to specify local, national, or international call (ISDN).
NLPID	Network Level Protocol Identifier, control field in frame header identifying encapsulated protocol.	**NPSI**	Network Packet Switching Interface, IBM software for packet connection to FEP (SNA).
NM	Network Management.		
NME	NM Element.	**NR**	Number Received, control field sequence, tells sender the NS that receiver expects in next frame (Layer 2).
NMOS	N-channel Metal Oxide Semiconductor, common IC type uses more power than CMOS.		
NMP	Network Management Protocol.	**NREN**	National Research and Education Network, U.S.
NMS	Network Management System.	**NRZ**	Non-Return to Zero, signal transitions from positive to negative without assuming 0 value. See also DMC, AMI.
NMVT	Network Management Vector Transport (SNA).		
NNI	Network-Network Interface, between two carriers or between carrier and private network (FR, ATM).	**NRZI**	NRZ Invert on ones, coding changes polarity to indicate '1' and remains unchanged for '0.'
NNI	Network-Node Interface, point to point interface between two switches for SDH, SONET, or B-ISDN network.	**ns**	Nanosecond, 10^{-9} second.
		NS	Number Sent, sequence number of frame in its control field; determined by sender.
NOS	Network Operating System.		
NPA	Numbering Plan Area, area code in phone number: NPA-NXX-5555.	**NSA**	National Security Agency.
		NSA	Non-Service Affecting, fault that does not interrupt transmission.

NSAP Network-layer Service Access Point, logical address of a 'user' within a protocol stack (ISDN).

NSC Network Service Center, for SDN.

NSDU Network Service Data Unit, basic packet passed by SCCP (SS7); also OSI.

NSFNET National Science Foundation Network.

NSP Network Services Part, reliable transport for signaling, MTP + SCCP.

NT-1 Network Termination 1, the first device on the CP end of the ISDN local loop (like the CSU).

NT-2 Network Termination 2, the second CP device, like the DSU (ISDN).

NTM NT Test Mode, BRI control bit.

NTN Network Terminal Number, address of terminal on data network, part of global address with DNIC (X.121).

NTSC National Television Standards Committee, group and format they defined for U.S. TV broadcasting.

NTT Nippon Telephone and Telegraph, the domestic phone company in Japan.

NUI Network/User Interface.

NV NetView, IBM's umbrella network management system.

NXX Generic indication of exchange in phone number: NPA-NXX-5555.

NYSERnet New York State Education and Research Network, part of NSFnet.

O

OAI Open Applications Interface, Intecom's PHI.

OA&M Operations, Administration, and Maintenance.

OAM&P Operations, Administration, Maintenance, & Provisioning, telco housekeeping.

OC-1 Optical Carrier level 1, SONET rate of 51.84 Mbit/s, matches STS-1.

OC-3 Optical Carrier level 3, SONET rate of 155.52 Mbit/s, matches STS-3.

OC-N Higher SONET levels, N times 51.84 Mbit/s.

OCR Office Channel Repeater, OCU.

OCU Office Channel Unit, "CSU" in the CO; also called OCR.

OCU-DP OCU-Data Port, channel bank plug I/O to 4-wire local loop and CSU on CP to provide DDS.

ODI Open Data-link Interface, driver interface, API for LAN cards.

OF Optical Fiber.

OLTP On Line Transaction Processing.

OMAP Operations Maintenance and Administration Part, upper layer 7 protocol in SS7.

ONA Open Network Architecture, FCC plan for equal access to public networks.

ONI   Operator Number Identification.

OOF   Out Of Frame, mux is searching for framing bit pattern.

OOS   Out Of Synchronization; multiplexers can't transmit data when OOS.

OPC   Origination signal transfer Point Code, address in SU of source of packet (SS7).

OPR   Optical Power Received, by a FO termination.

OPX   Off-Premises Extension, line from PBX to another site.

OR   Or, as in either/or, a logical device that outputs a 1 if any input is 1; a 0 only if all inputs are 0.

ORL   Optical Return Loss.

OS   Operating System, main software to run a CPU.

OSS   Operations and Support System, used by telco to provision, monitor, and maintain facilities.

OSF   Open Software Foundation.

OSI   Open Systems Interconnection, a 7-layer model for protocols defined by the ISO.

OSI/NMF   OSI Network Management Forum, standards group for NM protocols.

OSIone   Global organization to promote OSI standards.

OSI TP   OSI Transaction Processing, a protocol.

OSPF   Open Shortest Path First, standard routing protocol.

OSS   Operations (or Operational) Support System, used by telco to provision, monitor, and maintain facilities.

OTC   Operating Telephone Company, LEC.

OTC   Overseas Telephone Company, international carrier in Australia.

OTDR   Optical Time Domain Reflectometry (Reflectometer), method (tester) to locate breaks in OF.

OUI   Organizationally Unique Identifier, code for administrator of PIDs.

OW   Order Wire, DS-0 in overhead intended for voice path to support maintenance.

P

PA   Preamble, a period of usually steady signal ahead of a LAN frame, to set timing, reserve the cable, etc.

PABX   Private Automated Branch eXchange, electronic PBX.

PAD   Packet Assembler/ Disassembler, device to convert between packets (X.25, etc.) and sync or async data.

PAL   Programmable Array Logic, large semi-custom chip.

PAM   Pulse Amplitude Modulation; used within older channel banks and at 2B1Q ISDN U interface.

PANS   Peculiar and Novel Services, phone services that go beyond POTS: switched data, ISDN, etc.

PAP   Password Authentication Protocol, encrypts passwords for security of dial-in access.

PARIS   Packetized Automated Routing Integrated System, fast switch developed by IBM.

PBX   Private Branch eXchange, small phone switch inside a company, manual or automatic.

PC   Path Control, level 3 in SNA for network routing.

PCB   Printed Circuit Board.

PCI   Peripheral Component Interconnect, Intel's advanced bus for personal computers.

PCM   Page Counter Modulus, SMDS header field.

PCM   Pulse Code Modulation, the standard digital voice format at 64 kbit/s.

PCN   Personal Communications Network, second generation cellular system.

PCMCIA   Personal Computer Memory Card International Association.

PCR   Peak Cell Rate, traffic parameter applied per VC, VP, or channel (ATM).

PCR   Preventive Cyclic Retransmission, error correction procedure that repeats packets whenever link bandwidth is available (SS7).

PCR   Program Clock Reference, MPEG2 time stamp.

PCS   Personal Communications Service, low-power portable phones based on dense public networks of small cells.

PDG   Packet Data Group, 12 octets in FDDI frame (outside of WBCs) not assignable to circuit-switched connections.

PDH   Plesiochronous Digital Hierarchy, present multiplexing scheme from T-1 to T-3 and higher; contrast with SDH.

PDN   Public Data Network; usually packetized.

PDS   Premises Distribution System, the voice and data wiring inside a customer office.

PDU   Protocol Data Unit, information packet (ADDR, CTRL, INFO) passed at one level between different protocol stacks (OSI).

pel   Picture Element, the smallest portion of a graphic image encoded digitally.

P/F   Poll/Final, bit in control field of LLC frames to indicate receiver must acknowledge (P) or this is last frame (F) (Layer 2).

PFT Power Failure Transfer, protection switch.

PHF Packet Handling Facility, packet switch for X.25 or FR service (ISDN).

PHI PBX-Host Interface, generic term for link between voice switch and computer, c.f., SCAI.

PHY PHYsical, layer 1 of the OSI model.

PIC Polyethylene Insulated Cable, modern phone wire.

PID Protocol ID, codes (some allotted by CCITT) to identify specific proto-cols.

PIN Positive-Intrinsic-Negative, type of semi-conductor photo detec-tor.

PIU Path Information Unit, BIU plus the transmis-sion layer frame header (SNA).

PL Pad Length, number (0-3) of octets of 0s added to make Info field a mul-tiple of 4 octets (802.6).

PL Physical Layer, level 1 in OSI model.

PL Private Line, a dedicated leased line, not switched.

PLAR Private Line Automatic Ring-down; see ARD.

PLB Performance Loop Back, LB done at point of ESF performance function in CPE.

PLL Phase Locked Loop, electronic circuit that recovers clock timing from data.

PLP Packet Layer Protocol, at layer 3 like X.25.

PLS Physical Link Signaling, part of Layer 1 that encodes and decodes transmissions, e.g. Manchester coding (IEEE 802).

PMA Physical Medium Attachment, electrical driver for specific LAN cable in MAU, separated from PLS by AUI (802.3).

PMA Primary Market Area, metro area as served by MAN.

PMD Packet Mode Data, ISDN call type.

PMD Physical layer, Medium Dependent; a sublayer in layer 1 (below PLS) of LAN protocols; also PMA (802).

POF Plastic Optical Fiber, for short distances rather than glass for long haul.

POH Path OverHead, bytes in SDH for channels carried between switches over multiple lines and through DCCs .

POP Point Of Presence; end of IXC portion of long-distance line at central office (Tel).

POS Point of Sale.

POTS Plain Old Telephone Service, residential type analog service.

PPDU Presentation (layer) PDU (OSI).

ppm Parts Per Million, 1 ppm = 0.0001%.

PPP	Point to Point Protocol, non-proprietary multi-protocol serial interface for WAN links.
pps	Packets Per Second, switch capacity.
pps	Pulses Per Second, speed of rotary dialing dial pulses.
PRC	Primary Reference Clock, GPS-controlled rubidium oscillator used as stratum 1 source.
PRCA	Proportional Rate Control Algorithm.
PRA	Primary Rate Access, via PRI for ISDN.
PRBS	Pseudo-Random Bit Sequence, fixed bit pattern, for testing, that looks random but repeats.
PRI	Primary Rate Interface; 23B+D (T-1) or 30B+D (CEPT).
PRM	Protocol Reference Model.
PROM	Programmable Read Only Memory; non-volatile type chip.
PRS	Primary Rate Source, stratum 1 clock.
PS	Power Status, 2-bit control field at BRI.
PS	Presentation Services, level 6 of SNA.
PSC	Public Service Commission, telecom regulator in many states, also PUC.
PSDN	Public Switched Data Network, national collection of interconnected PSDSs.

PSDS	Public Switched Digital Service, generic switched 56K intra-LATA service.
PSI	Primary Subnet Identifier, part of address in network level header (MAP).
PSN	Packet Switched Network.
PSN	Public Switched Network.
PSPDN	Packet Switched Public Data Network.
PSTN	Public Switched Telephone Network, the telco-owned dial-up network.
PTAT	Private Trans-Atlantic Telephone, cable from US to U.K., Ireland, and Bermuda.
PTE	Path Terminating Equipment, SONET nodes on ends of logical connections.
PTI	Payload Type Identifier, control field in ATM header.
PTT	Postal, Telephone, and Telegraph authority; a monopoly in most countries.
PU	Physical Unit, SNA protocol stack that provides services to a node and to less intelligent devices attached to it.
PU2	Cluster controller or end system.
PU4	Front End Processor.
PUB	AT&T technical PUBlication, Bell System de facto standard, most from before divestiture.

PUC   Public Utilities Commission, state body that regulates telephones, also PSC.

PVC   Permanent Virtual Circuit (Connection), assigned connection over a packet, frame, or cell network, not switchable by user.

PVN   Private Virtual Network, VPN.

PWB   Printed Wiring Board, PCB.

Q

QLLC   Qualified Logical Link Control, a frame format (SNA).

QOS   Quality of Service.

QPSX   Queued Packet Synchronous eXchange, old name for DQDB; QPSX Systems Inc. originated it in Australia.

Q.921   CCITT recommendation for level 2 protocol in signaling system 7.

Q.931   CCITT recommendation for level 3 protocol in signaling system 7.

R

R   Interface reference point in the ISDN model to pre-ISDN phone or terminal.

R   Red alarm bit in synch byte (TS 24) of T1DM (Tel).

R   Reserved, bit or field in frame not yet standardized, not to be used.

R   Ring, one of the conductors in a standard twisted pair, 2-wire local loop (the one connected to the 'ring,' the second part of a phone plug) or the DTE-to-DCE side of a 4-wire interface.

R_1   Ring, or R lead of the DCE-to-DTE pair in a 4-wire interface.

RACE   Research for Advanced Communications in Europe, program to develop broadband.

RACF   Remote Access Control Facility, security program (SNA).

RAI   Remote Alarm Indication, (yellow alarm) repeating pattern of 8 ones and 8 zeros in EOC of ESF T-1 line (also in ATM).

RAID   Redundant Array of Inexpensive (Independent) Disks.

RAM   Random Access Memory; volatile chip.

RARE   Reseaux Associes pour la Recherche Europeene, European Organization of Research Networks.

RARP   Reverse ARP, Internet protocol to let diskless workstation learn its IP address from a server (TCP/IP).

RBB   Residential BroadBand.

RBHC   Regional Bell Holding Company, one of the seven "baby Bells."

RBOC Regional Bell Operating Company, one of about 22 local telephone companies formerly part of Bell System.

RBS Robbed Bit Signaling, how channel bank places dial digits and on/off hook supervision in same channel as PCM voice encoding.

RD Receive Data, lead on electrical interface.

RD Request Disconnect, secondary station unnumbered frame asking primary station for DISC (Layer 2).

RDA Remote Database Access, service element (OSI).

RDT Remote Digital Terminal, advanced channel bank functionality on fiber or copper loop.

REJ Reject, S-format LLC frame acknowledges received data units while requesting retransmission from specific errored frame (Layer 2).

REL RELease, signaling packet on disconnect (SS7).

RELC Release Complete, packet to acknowledge disconnect (DSS1 and SS7).

RELP Residually-Excited Linear Predictive Coding, voice encoding scheme (8-16 kbit/s).

REQ Request (OSI).

RF Radio Frequency.

RFC Request For Comment, documents that are modified then adopted by IETF as Internet standards.

RFH Remote Frame Handler, FR switch or network accessed over CS links.

RFI Radio Frequency Interference.

RFT Remote Fiber Terminal, equivalent to SLC96.

RGB Red Green Blue, primary colors encoded in video signal.

RH Request/response Header, 3 bytes added to user data in format for first upper layer frame (SNA).

RHC Regional Holding Company, one of the 7 telco groups split from AT&T in 1984, see RBOC.

RI Ring Indicator, digital lead on modem tells DTE when call comes in (phone rings).

RI Routing Indicator, bit in LAN packet header to distinguish transparent- from source-routed packets.

RIM Request Initialization Mode, layer 2 supervisory frame.

RIP Routing Information Protocol, method for routers to learn LAN topology (TCP/IP).

RISC Reduced Instruction Set Computer.

RJ Registered Jack, connector for UNI; RJ11 is standard phone, RJ45 for DDS and terminal, RJ48 for T-1.

RJE Remote Job Entry, one form of BSC.

RL Ring Latency, time for empty token to traverse full ring with no load (FDDI).

RM Reference Model.

RM Resource Management, type of overhead ATM cell.

RMN Remote Multiplexing Node.

RN Redirecting Number, DN of party that forwarded a call via the network (ISDN).

RNR Receiver Not Ready, S-format LLC frame acknowledges received data units but stops sender temporarily (Layer 2 HDLC).

RO Receive Only.

ROLC Routing Over Large Clouds, IETF study.

ROM Read Only Memory; non-volatile chip.

ROSE Remote Operation Service Element (OSI).

RPOA Recognized Private Operating Agency, X.25 interexchange carrier (ISDN).

RR Receive Ready, S-format LLC frame acknowledges received data units and shows ability to receive more (Layer 2).

RS Radiocommunications Sector, part of ITU, 1993 successor to CCIR.

RSET Reset, layer 2 supervisory frame to zero counters.

RSL Request and Status Link, same as PHI or SCAI.

RSP Response (OSI).

RSU Remote Switch Unit, multiplexing equipment outside CO that serves a CSA.

RT Remote Terminal, CPE end of multiplexed access loop.

RTS Request To Send; lead on terminal interface.

RTU Remote Terminal (Test) Unit.

RU Request/response Unit, unframed block of up to 256 bytes of user data (SNA).

RVI Reverse Interrupt, positive ACK that lets station take control of a BSC line.

RZ Return to Zero; signal pauses at zero voltage between each pulse, when making zero crossings.

S

S Status, signaling bit in CMI.

S ISDN interface point between TA and NT-2.

S Supervisory frame, commands at LLC level: RR, RNR, REJ, SREJ (Layer 2).

S0 European notation for BRI.

S2 European notation for PRI (30B+D).

SA Source Address, field in frame header (802).

SA Synchronous Allocation, time allocated to FDDI station for sending sync frames (802.6).

SAA Systems Application Architecture, compatibility scheme for communications among IBM computers.

SABM Set Asynchronous Balanced Mode, connection request between HDLC controllers or LLC entities (Layer 2).

SABME SABM Extended, uses optional 16-bit control fields.

SAFER Split Access Flexible Egress Routing, service at one site from two toll offices over separate T-1 loops (AT&T).

SAI S/T Activity Indicator, BRI control bit.

SAP Service Access Point, logical address of a session within a physical station, part of a header address at an interface between sublayers (802).

SAP Service Advertising Protocol, periodic broadcast by LAN device (Netware).

SAPI Service Access Point Identifier, part of address between layers in protocol stack; e.g., subfield in first octet of LAP-D address.

SAR Segmentation And Reassembly, protocol layer that divides packets into cells.

SAR-PDU SAR Protocol Data Unit, segment of CS-PDU with additional header and possibly a trailer (e.g., a cell in ATM).

SARM Set Asynchronous Response Mode, unnumbered frame connection request (layer 2 HDLC).

SARME SARM Extended, uses optional 16-bit control field.

SARTS Special Access Remote Test System, the way telcos test leased lines.

SAS Single-Attached Station, FDDI node linked to network by 2 optical fibers (vs. DAS).

SB Signal Battery, second lead to balance M lead in E&M circuit.

SCA Selected Call Appearance, signaling message from key set to SPCS.

SCADA Supervisory Control and Data Acquisition, in nets for oil and gas producers.

SCAI Switch-to-Computer Applications Interface, link between host CPU and voice switch to integrate applications; also PHI and RSL.

SCCP Signaling Connection Control Part, upper layer 3 protocol (SS7).

SCIL Switch Computer Interface Link, PHI by Aristacom.

SCP   Service Control Point, CPU linked to SS7 that supports carrier services (800, LIDB, CLASS).

SCPC   Single Channel Per Carrier, analog satellite technology (telephony).

SCR   Sustainable Cell Rate, traffic parameter (ATM).

SD   Starting Delimiter, unique symbol to mark start of LAN frame (JK in FDDI, HDLC flag, etc.).

S/D   Signal to Distortion ratio.

SDDN   Software Defined Data Network, virtual private network built on public data net.

SDH   Synchronous Digital Hierarchy, digital multi-plexing plan where all levels are synched to same master clock,

SDLC   Synchronous Data Link Control; a half-duplex IBM protocol based on HDLC.

SDM   Subrate Digital (Data) Multiplexing, a DDS service to put multiple low-speed channels in a DS-0.

SDN   Software Defined Net-work.

SDS   Switched Digital Service, generic term for carrier function.

SDU   Service Data Unit, information packet or segment passed down to become the payload of the adjacent lower layer in a protocol stack.

SEND   clear to Send, signaling bit in CMI.

SES   Severely Errored Second, interval when BER exceeds 10-3, >319 CRC errors in ESF, frame slip, or alarm is present.

SF   Single Frequency; form of on/off-hook analog signaling within telcos.

SF   Subfield (SNA).

SF   Super Frame, 12 T-1 frames.

SFET   Synchronous Frequency Encoding Technique, a way to send precise isoc clocking rate as a delta from system clock.

SG   Signal Ground, second lead to balance E lead in E&M signal circuit.

SG   Study Group, committee of ITU.

SHR   Self-Healing Ring, topology can survive one failure in line or node (802.6, etc.).

SI   Sequenced Information, LAP-D frame type.

SIF   Signaling Information Field, payload of a signaling packet or MSU (SS7).

SIM   Set Initialization Mode, layer 2 supervisory frame.

SIO   Service Information Octet, field in MSU used to identify individual users (SS7).

SIP   SMDS Interface Protocol.

SIPO   Signaling Indication Processor Outage, alarm on failure of processor that receives signaling packets ("indications") (SS7).

SIR   Sustained Information Rate, average through-put; basis for SMDS access class.

SIT   Special Information Tone, audible signal (often three rising notes) preceding an announce-ment by the network to a caller.

SITA   Societe Internationale de Telecommunications Aeronautiques, operator of worldwide airline network.

SIVR   Speaker Independent Voice Recognition.

SLC   Subscriber Loop Carrier, usually digital loop system.

SLIC   Subscriber Line Interface Card (Circuit), on a switch.

SLIP   Serial Line Internet Protocol, older PPP for IP only.

SLS   Signaling Link Selection, field in routing label of SU that keeps related packets on same path to preserve delivery order (SS7).

SMAP   Systems Management Application Process, all the functions at layer 7 and above to monitor and control the network (SS7).

SMB   Server Message Block, a LAN client-server protocol.

SMDR   Station Message Detail Recording, keeping list of all calls from each phone, usually by PBX or computer.

SMDS   Switched Multi-megabit Data Service, offered on a MAN by a carrier; service mark of Bellcore.

SME   Subject Matter Expert.

SMF   Single Mode Fiber, thin strand that supports only one transmission mode for low dispersion of optical waves.

SMP   Simple Management Protocol, newer and more robust than SNMP, but nearly as successful.

SMR   Specialized Mobile Radio, for fleet manage-ment and dispatching.

SMT   Station ManagemenT, NMS for FDDI.

SMTP   Simple Mail Transfer Protocol (TCP/IP).

S/MUX Workstation software to allow UNIX daemons to talk to SNMP manager station.

SN   Sequence Number, transmission order of frames or cells within channel or logical connection.

SNA   SDH Network Aspects, evolving standards for VC payloads and net-work management (SDH).

SNA   Systems Network Architecture, IBM's data communication scheme.

SNADS SNA Distribution Services, communication architecture for electronic mail and other applica-tions.

SNAP	Sub-Network Access Protocol, identifies encapsulated protocol and user (802.1, ATM).	**SPDU**	Session (layer) PDU (OSI).
SNI	Subscriber-Network Interface, the demark point.	**SPE**	Synchronous Payload Envelope, data area in SONET/STS/SDH format, with POH.
SNMP	Simple Network Management Protocol, started in TCP/IP, but extending to many LAN devices (Layer 4-5).	**SPF**	Shortest Path First, LAN router protocol that minimizes some measure (delay) and not just "hops" between nodes.
SNR	Signal to Noise Ratio, in dB.	**SPID**	Service Profile IDentifier, DN or DN plus unique extension (ISDN in N.A.).
SNRM	Set Normal Response Mode, unnumbered command frame (layer 2).	**SQPA**	Software Quality Program Analysis, Bellcore process to evaluate vendors.
SNRME	SNRM Extended, uses optional 16-bit control field.	**SREJ**	Selective REJ, layer 2 frame that requests retransmission of one specific I frame.
SO	Serving Office, central office where IXC has POP.		
SOH	Section OverHead, bytes in SDH for channels carried through repeaters between line terminations like DCC or switch.	**SRL**	Singing Return Loss.
		SRT	Source Routing Transparent, variation of source routing combined with spanning tree algorithm for bridging (802).
SOH	Start of Header, control byte in BSC.	**SRTS**	Synchronous Residual Time Stamp.
SOHO	Small Office Home Office.	**SS7**	Signaling System 7, replaced CCIS or SS6 in ISDN.
SONET	Synchronous Optical Network.	**SSA**	Systems Applications Architecture, SNA plan to allow programs on different computers to communicate.
SPAG	Standards Promotion and Applications Group, has same function as COS.		
SPCS	Stored Program Controlled Switch, CO switch (analog or digital) controlled by a computer.	**SSAP**	Source Service Access Point, field in LLC frame header to identify the sending session within a physical station (802).

SSCF   Service Specific Coordination Function, maps SSCOP functions to lower layer (Q.2130, ATM).

SSCP   System Services Control Point, host software that controls SNA network.

SSCS   Service Specific Convergence Sublayer (ATM).

SSM   Single Segment Message, frame short enough to be carried in one cell.

SSN   SubSystem Number, local address of SCCP user (SS7).

SSP   Service Switching Point (ISDN).

ST   Stream, network layer protocol for very high speed connections.

STDM   Statistical Time Division Multiplexer.

STE   Section Terminating Equipment, SONET repeater.

STEP   Speech and Telephony Environment for Programmers, Wang's PHI.

STM   Synchronous Transfer Mode, one of several possible formats for SONET and BISDN.

STM-1   Synchronous Transport Module-1, smallest SDH bandwidth; = 155.52 Mbit/s, STM-n = n x 155.52 Mbit/s.

STP   Shielded Twisted Pair, telephone cable with additional shielding for high speed data and LANs.

STP   Signal Transfer Point, packet switch for SS7.

STS-1   Synchronous Transport Signal, level 1; electrical equivalent of OC-1, 51.84 Mbit/s.

STS-N   Signal in STS format at N x 51.84 Mbit/s.

STSX-n   Interface for cross-connect of STS-n signal that defines STS-n.

STX   Start of Text, control byte in BSC.

SU   Signaling Unit, layer 3 packet (SS7).

SV   Subvector, part of NMVT (SNA).

SVC   Switched Virtual Circuit (Connection), temporary logical connection in a packet/frame network.

SVD   Simultaneous Voice and Data.

SWG   SubWorking Group, part of a technical committee or forum.

SWIFT   SWItched Fractional T-1, telco service defined by Bellcore, includes full T-1.

SWIFT   Society for Worldwide Interbank Financial Telecommunications, global funds transfer network of 2000 banks.

SW56   Switched 56 kilobit/s, digital dial up service.

SYN   Synchronization character, 16h ASCII.

sync   Synchronous.

SYNTRAN   Synchronous Transmission, byte aligned format for an electrical DS-3 interface; obsolete.

T

T Interface between NT-1 and NT-2 (ISDN).

T Non-data character in 4B/5B coding, ending delimiter (802.6).

T Measurement interval, seconds, = Bc/CIR.

T Tip, one of the conductors in a standard twisted pair, 2-wire local loop (the wire connected to the 'tip' of a phone plug) or one of the DTE-to-DCE pair of a 4-wire interface.

T Transparent, no robbed bit signaling in D4/ESF format.

T-1 Transmission at DS-1, 1.544 Mbit/s.

T1 The standards committee responsible for transmission issues in US, corresponds to ETSI (Europe) and the Telecommunications Technology Committee (Japan).

T_1 Tip or T lead of the DCE-to-DTE pair in a 4-wire interface.

T1DM T-1 Data Multiplexer, brings DS-0Bs together on a DS-1 (Tel).

T1D1 TSC of T1 for BRI U interface.

T1E1 TSC of T1 for SNI.

T1M1 TSC of T1 for NMS and OSS.

T1Q1 TSC of T1 for ADPCM, voice compression, etc.

T1S1 TSC of T1 for ISDN bearer services.

T1X1 TSC of T1 for SONET and SS7.

TA Technical Advisory, a Bellcore standard in draft form, before becoming a TR.

TA Terminal Adapter, matches ISDN formats (S/T) to existing interfaces (R) like V.35, RS-232.

TABS Telemetry Asynchronous Block Serial, M/S packet protocol used to control network elements and get ESF stats.

TAC Technical Assistance Center, network help desk.

TAPS Test and Acceptance Procedures, telco document for equipment installation and set up.

TASI Time Assigned Speech Interpolation; analog voice compression comparable to DSI and statistical multiplexing of data.

TAT Trans-Atlantic Telephone, applied to cables, as TAT-8.

TAXI Transparent Asynchronous Xmit/receive Interface, 100 Mbit/s interface to ATM switch.

TBD To Be Determined, appears often in unfinished technical standards.

TC Terminating Channel; local loop.

TC Transport Connection.

TC Transmission Control, level 4 in SNA.

TC   Trunk Conditioning, insertion of various signaling bits in A and B positions of DS-0 during carrier failure alarm condition.

TCA   TeleCommunications Association.

TCA   Threshold Crossing Alert, alarm that a monitored statistic has exceeded preset value.

TCAP   Transaction Capabilities Application Part, lower layer 7 of SS7.

TCC   Telephone Country Code, part of dialing plan.

TCP/IP   Transmission Control Protocol (connection oriented with error correction) often runs on Internet Protocol (a connectionless datagram service).

TD   Transmit Data.

TDD   Telecom Device for the Deaf, Teletype machine or terminal with modem for dial-up access.

TDM   Time Division Multiplexing (or Multiplexer).

TDMA   Time Division Multiple Access, stations take turns sending in bursts, via satellite or LAN.

TDS   Terrestrial Digital Service, MCI's T-1 and DS-3 service.

TDSAI   Transit Delay Selection And Indication, way to negotiate delay across X.25 bearer service (ISDN).

TE   Terminal Equipment, any user device (phone, fax, computer) on ISDN service; TE1 supports native ISDN or B-ISDN formats (S/T interface); TE2 needs a TA.

TEI   Terminal Endpoint Identifier, subfield in second octet of LAP-D address field (ISDN).

TEST   Test command, LLC UI frame to create loopback (Layer 2).

TG   Transmission Group, one or more links between adjacent nodes (SNA).

TH   Transmission Header, 2 bytes in framing format for layer 4 protocol (SNA).

TIA   Telecommunications Industry Association, successor to EIA, sets some comms standards.

TIFF   Tagged Image File Format, for graphics files.

TIRKS   Trunk Inventory Record Keeping System, telco computer to track lines.

TIU   Terminal Interface Unit, CSU/DSU or NT1 for Switched 56K service that handles dialing.

TLA   Three Letter Acronym.

TLI   Transport Level Interface, for UNIX.

TL1   Transaction Language 1, to control network elements (TR482); CCITT's form of MML.

TLP	Transmission Level Point, related to gain (or loss) in voice channel; measured power - TLP at that point = power at 0 TLP site.	**TPEX**	Twisted Pair Ethernet Transceiver.
		TP-N	Transport Protocol of Class N (N=0 to 4), OSI layer 4.
TM	Traffic Management (ATM).	**TP-0**	Connectionless TP (ISO 8602).
TMN	Telecommunications Management Network, a support network to run a SONET network.	**TP-4**	Connection oriented TP (ISO 8073).
		TPDU	Transport Protocol Data Unit (OSI).
TMS	Timing Monitoring System.	**TPF**	Transaction Processing Facility, IBM host software for OLTP.
TN	TelNet, remote ASCII terminal emulation (TCP/IP).	**TPSE**	Transport Processing Service Element (OSI).
TN	Transit Network, IEC (ISDN).	**TR**	Technical Reference, a final Bellcore standard.
TN3270	Remote emulation of IBM 3270 terminal.	**TR**	Token Ring, a form of LAN.
TO	Transmit Only; audio plug for a channel bank without signaling.	**TS**	Time Slot, DS-0 channel in T-1, PRI, etc.
TOA	Type of Address, 1-bit field to indicate X.121 or not (X.25).	**TS**	Telecommunications Standardization bureau, formed by ITU in 1993 from merger of CCITT and CCIR.
TON	Type Of Number, part of ISDN address indicating national, international, etc.	**TS**	Time Stamp, information used to correlate CBR clock rates at both ends of a connection (ATM).
TOP	Technical and Office Protocol; for LANs.	**TS**	Transaction Services, top level (7) of SNA protocol stack, on top of LU 6.2.
TOPS	Task Oriented Procedures, telco document for equipment operation and maintenance.	**TS**	Transport Service (OSI).
		TS	Transport Stream.
TOS	Type Of Service, connection attribute used to select route in LAN (SPF).	**TSB**	Telecommunications Standardization Bureau, formed by ITU in 1993 from merger of CCITT and CCIR.
TP	Transaction Processing, work of a terminal on-line with a host computer.	**TSC**	Technical Subcommittee, for standards setting.

TSDU Transport Service Data Unit (OSI).

TSI Time Slot Interchange(r); method (device) for temporarily storing data bytes so they can be sent in a different order than received; a way to switch voice or data among DS-0s (DACS).

TSS Telecommunications Standardization Sector, a variant on TSB.

TSY Technology Systems, Bellcore group renamed Network Technology (NWT).

TTC Telecommunications Technology Committee, Japanese standards body.

TTL Transistor-Transistor Logic; signals between chips.

TTRT Target Token-Rotation Time, expected or allowed period for token to circulate once around ring (802.4, 802.6).

TTY Teletypewriter.

TU Tributary Unit, virtual container plus path overhead (SDH).

TUC Total User Cells, count kept per VC while monitoring, field in OAM cell.

TUG TU Group, one or more TUs multiplexed into a larger VC (SDH).

TUP Telephone Users Part, ISDN signaling based on MTP without SCCP, used outside N.A. only.

TWX TeletypeWriter Exchange, switched service (originally Western Union) separate from Telex.

U

u English transliteration of Greek mu (μ), for micro or millionth; prefix in abbreviation of units like us, um.

U Interface between CO and CP for ISDN.

U Unnumbered format, command frames, same as UI (Layer 2).

U Rack Unit, vertical space of 1.75 inch.

UA Unnumbered Acknowledgement, LLC frame to accept connection request (Layer 2).

UART Universal Async Receiver Transmitter, interface chip for serial async port.

UAS Unavailable Second, when BER of line has exceeded 10-3 for 10 consecutive seconds until next AVS start.

UBR Unspecified Bit Rate, service with no bandwidth reservation (ATM).

UDLC Universal Data Link Control, Sperry Univac's HDLC

UDP/IP Universal Data Protocol or User Datagram Protocol, Internet Protocol; UDP is a transport layer without error correction.

uF   Microfarad, one mil-
           lionth of the unit of
           capacitance.

UI   Unnumbered Informa-
           tion, frame at LLC level
           whose control field
           begins with 11: XID,
           TEST, SABME, UA, DM,
           DISC, FRMR (802).

UID   User-Interactive Data,
           circuit mode digital
           transport (ISDN).

ULP   Upper Layer Protocol.

UNI   User-Network Interface,
           demark point of ATM,
           SDH, FR, and B-ISDN at
           customer premises.

UNMA   Unified Network Man-
           agement Architecture;
           AT&T's umbrella soft-
           ware system.

UNR   Uncontrolled Not Ready,
           signaling bit in CMI.

UOA   U-interface Only Activa-
           tion, BRI control bit.

UP   Unnumbered Poll,
           command frame (Layer
           2).

U-Plane User Plane, bearer circuit
           for customer informa-
           tion, controlled by
           C-Plane.

UPS   Uninterruptable Power
           Supply.

us   Microsecond; 10^{-6}
           second.

USART Universal Sync/Async
           Receiver Transmitter,
           interface chip for sync
           and async data I/O.

USAT   Ultra-Small Aperture
           SATellite; uses ground
           station antenna less than
           1 m diameter.

USOC   Universal Service Order
           Code.

UTC   Universal Coordinated
           Time, the ultimate global
           time reference.

UTP   Unshielded Twisted Pair,
           copper wire.

UUSCC User to User Signaling
           with Call Control, ISDN
           feature that passes user
           data with some signaling
           messages.

V

V.25bis Dialing command
           protocol for modems,
           CSUs, etc.

V.35   CCITT recommendation
           for 48 kbit/s modem that
           defined a data interface;
           replaced by V.11 (electri-
           cal) and EIA-530 (me-
           chanical and pinout on
           DB-25).

VAN   Value Added Network;
           generally a packet
           switched network with
           access to data bases,
           protocol conversion, etc.

VBD   Voice Band Data, ISDN
           terminal mode that may
           include a modem or fax.

VBR   Variable Bit Rate,
           packetized bandwidth
           on demand, not dedi-
           cated (ATM).

VC   Virtual Container, a cell
           of bytes carrying a
           slower channel to define
           a path in SDH; VC-n
           corresponds to DS-n,
           n = 1 to 4.

VC   Virtual Circuit (Channel), logical connection in packet network so net can transfer data between two ports.

VCC   Virtual Circuit (Channel) Connection; between terminals (SMDS, SONET, ATM).

VCI   Virtual Circuit (Channel) Identifier; part of a packet/frame/cell address in header (802.6, ATM).

VDT   Video Display Terminal, often applied to any type of "tube" or PC.

VESA   Video Electronics Standards Assoc., defined the VESA-bus for personal computers.

VF   Voice Frequency, 300-3300 Hz or up to 4000 Hz.

VG   Voice Grade; related to the common analog phone line.

VGPL   Voice Grade Private Line, an analog line.

VHF   Very High Frequency, radio band from 30 to 300 MHz.

VLAN   Virtual LAN, term for logical LAN connectivity based on need rather than physical connection.

VLSI   Very Large Scale Integration, putting thousands of transistors on a single chip.

VMTP   Versatile Message Transport Protocol, designed at Stanford to replace TCP and TP4 in high-speed networks; didn't.

VNL   Via Net Loss, related to TLP.

VOD   Video on Demand.

VOX   Voice Activation, in voice over FR, silence suppression by not sending frames when audio level is below threshold.

VP   Virtual Path, for many VCCs between concentrators (ATM).

VPC   VP Connection.

VPI   Virtual Path Identifier, VCI in ETSI version of ATM; applies to bundle of VCCs between same end points (ATM).

vPOTS   Very Plain Old Telephone Service; no switching; ARD etc.

VPC   Virtual Path Connection; between switches (SONET).

VPN   Virtual Private Network, logical association of many user sites into CUG on PSTN.

VPN   Virtual Private Network, closed user group constructed on a public data network for LAN traffic; may use PVCs or encryption.

VQC   Vector Quantizing Code; a voice compression technique that runs at 32 and 16 kbit/s.

VQL   Variable Quantizing Level; voice encoding method.

VR   Receive state Variable, value in register at receiver indicating next NS expected (Layer 2).

VR   Virtual Route (SNA).

VRU   Voice Response Unit, automated way to deliver information and accept DTMF inputs.

V_s Send state Variable, value in register of sender of NS in last frame sent (layer 2).

VSAT   Very Small Aperture Terminal, satellite dish under 1 m.

VSELP   Vector-Sum Excited Linear Prediction, compression algorithm used in some digital cellular systems.

VT   Virtual Tributary, logical channel made up of a sequence of cells within SONET or similar facility.

VTAM   Virtual Telecommunications Access Method, SNA protocol and host communications program.

VTE   Virtual Tributary Envelope, the real payload plus path overhead within a VT (SONET).

V.35   Former CCITT recommendation for a modem with a 48 kbit/s interface on a large 44-pin connector, being replaced by EIA-530 pinout on DB-25.

W

WACK   Wait before transmit positive Acknowledgment, control sequence of DLE plus second character (30 ASCII, 6B EBCDIC).

WAN   Wide Area Network, the T-1, T-3, or broadband backbone that covers a large geographical area.

WATS   Wide Area Telephone Service.

WBC   WideBand Channel, one of 16 FDDI subframes of 6.144 Mbit/s assignable to packet or circuit connections.

W-DCS   Wideband Digital Cross-connect System, 3/1 DACS for OC-1, STS-1, DS-3, and below, including T-1 (see B-DCS).

WDM   Wavelength Division Multiplexing, 2 or more colors of light on 1 fiber.

X

X   X class central office switch is in HPR net but not linked to NP.

X   Any digit, 0-9.

X.25   CCITT recommendation defining Level 3 protocol to access a packet switched network.

XGMON   X-Windows (based) Graphics Monitor, IBM net management software for SNMP.

XID   Exchange Identification, type of UI command to exchange parameters between LLC entities (layer 2).

XNS   Xerox Network Services, a LAN protocol stack.

X-off   Transmit Off, ASCII character from receiver to stop sender.

X-on   Transmit On, ok to resume sending.

XTP   eXpress Transfer Protocol, a simplified low-processing protocol proposed for broadband networks.

Y Z

Y   Yellow alarm control bit in sync byte (TS 24) of T1DM, Y=0 indicates alarm.

Z   Impedance, nominal 600 ohm analog interface may be closer to complex value of 900 ohms R + 2 uF C.

ZBTSI   Zero Byte Time Slot Interchange, process to maintain 1's density.

Numeric

1Base5   1 Mbit/s BASEband signaling good for 500 m, STARLAN standard (802.3).

2B+D   Two Bearer plus a Data channel, format for ISDN basic rate access.

2B1Q   2 Binary 1 Quaternary, line code for BRI at U reference point.

2W   2-Wire, analog interface with send and receive on same pair of wires.

4B/5B   Coding that substitutes 5 bits for each 4 bits of data, leaving extra codes for commands (802.6 & FDDI). See also DMC, NRZ, AMI.

4W   4-Wire, analog or digital interface with receive and send on separate wire pairs.

5E8   Software release for 5ESS that supports National ISDN-1.

5ESS   Trademark for class 5 (end office) ESS made by AT&T.

10Base5   10 Mbit/s BASEband signaling good for 500 m, LAN definition of Ethernet (802.3).

10BaseT   10 Mbit/s BASEband signaling over twisted pair, Ethernet (802.3).

23B+D   23 Bearer Plus a Data channel, ISDN primary rate T-1 format.

30B+D   30 Bearer channels plus a Data channel, ISDN primary rate E-1 format.

800   Area code for phone service where called party pays the carrier for the call (Free Phone).

802.x   IEEE standards for LAN protocols.

802.1   Spanning tree algorithm implemented in bridges.

802.3   Ethernet.

802.4   Token Bus architecture for MAP LAN.

802.5   Token Ring, with source routing.

802.6   Distributed Queue, Dual Bus MAN.

802.10   LAN security.

802.11   Spread spectrum local radio for LANs.

900   Area code for phone service where calling party is charged for the call plus a fee that the carrier pays to called party.

54016   Specification for ESF.

62411   Basic description of T-1
             service and interface.

New Additions

You are invited to send your own
list to the author for inclusion in a
future edition.

Appendix

B

Annotated Bibliography

These are the main resource documents for this book and the primary references for ISDN. It is not intended to be exhaustive—there are many other books (and an enormous number of magazine articles) on the topic. More publications appear each month. Fortunately, most current literature is on applications, rather than changes to standards. However, for equipment design, ensure you have the latest version of a standard—request the status of a particular document from its publisher.

American National Standards Institute

T1.601-1992 **ISDN Basic Access Interface for Use on Metallic Loops for Application on the Network Side of the NT–Layer 1 Specification.** This is the U interface. It is cited in Bellcore documents which generally reproduce little of what is in this ANSI standard.

T1.605-1991 **ISDN Basic Access Interface for S and T Reference Points (Layer 1 Specification).** Defines physical and electrical properties of the S/T interface specified in I.430 and described in I.412.

International Telecommunications Union

ITU-T carried over from CCITT a practice of giving a document a new identification number in a different letter series when one working group wanted to cite the work of another working group. Thus the telephone numbering plan, E.164, is called I.331 by the ISDN working group. Avoid buying a document whose content is merely a reference to another title.

CCITT worked in 4-year cycles, ending with the publication of a book (many volumes) of revised recommendations (standards). The binding color was changed with each edition: the last one was the Blue Book in 1988. Since it became the ITU, publication has been more rapid, with individual documents issued when ready.

E.164 **Numbering Plan for the ISDN Era;** also called I.331.

I.331 **Numbering Plan for the ISDN Area.** Addressing. E.164.

I.412 **ISDN User-Network Interfaces; Interface Structures and Access Capabilities.** Defines type of channels and how they are combined on physical interfaces.

I.430 **ISDN Basic User-Network Interfaces—Layer 1 Specification;** Blue Book, 1988. For the S and T interfaces; became ANSI's T1.605.

I.431 **Primary Rate User-Network Interface—Layer 1 Specification;** Blue Book, 1988. Describes the electrical signals, framing, and usage for T-1 and E-1 PRIs between NT and TE.

I.440 **ISDN UNI Data Link Layer—General Aspects.** See Q.920.

Q.921 **ISDN UNI—Data Link Layer Specification.** Based on HDLC, with distinctions in the address field for SAPI and TEI. (I.441)

Q.931 **ISDN User-Network Interface Layer 3 Specifications,** CCITT, 1988 Blue Book. This is the original source for DSS-1 signaling in the D channel between customer premises and the serving central office.

V.110 **Support of DTEs with V-Series Type Interfaces by an ISDN.** Format for 56 kbit/s traffic in a 64 kbit/s channel. (I.463)

V.120 **Support by an ISDN of DTE with V-Series Type Interfaces with Provision for Statistical Multiplexing.** This adds the ability to address multiple users while rate adapting. (I.465)

Bellcore (renamed Telcordia)

Building a library of Bellcore reference materials was a never-ending job. One of Bellcore's hopes, it seemed, was to sell a new edition of every document every year to every reader—at a cost of at least $100 per document (certain sets cost more than $6,000). Telcordia seems to have a more stable document list, but the top prices now exceed $15,000 per set.

The organization, content, and numbering scheme for their documents also changed every few years, necessitating acquisition of new sets to be able to look up current cross references. While the ISDN specification itself is unlikely to change much from this point on, designers of equipment are well advised to use the latest editions from Telcordia to catch the additions and modifications being made as National ISDN 2 is deployed.

TA-NWT-001268 **ISDN PRI Call Control Switching and Signaling Generic Requirements for Class II Equipment;** Issue 1, March 1991. Based on Q.931; limited to a subset of most functions and does not include all ITU-T features.

SR-NWT-001937 **National ISDN-1;** Issue 1, February 1991. Summarizes the features supported in NI-1 and lists the documents that contain the detailed specifications for each feature.

SR-NWT-2120 **National ISDN-2;** Issue 1, May 1992, and Revision 1, June 1993. Same function as SR-1937, but more of it: 33 pages of feature descriptions.

SR-NWT-002343   **ISDN Primary Rate Interface Generic Guidelines for CPE,** Issue 1, June 1993. Major source, as it reflects both the CPE and network functions. Includes recent updates to ANSI documents.

SR-NWT-002661   **National ISDN Generic Guidelines for ISDN Terminal Equipment on Basic Rate Interfaces,** Issue 1 Rev. 1, Aug. 1993.

SR-3480   **Procedures for Performing an ISDN Ordering Code Translations Review;** Issue 1, May 1995, and Revision 1, August 1995. This document, new in 1995, summarizes the process to test and register sets of translations (switch commands) to implement specific capability sets on BRI lines. Lacks any basic explanations or even a glossary for the dozens of unexplained acronyms; basically a cookbook for switch technicians. Consult Telcordia for current procedures.

SR-NWT-002661   **National ISDN Generic Guidelines for ISDN Terminal Equipment on Basic Rate Interfaces.**   Includes excerpts from ANSI and other documents to describe the present state of the specification for S/T and U points of the BRI for; fairly complete—over 1 inch thick.

World Wide Web

Among the truly astounding volume of information offered on the Web is quite a lot on ISDN.

http://alumni.caltech.edu/~dank/isdn/

Dan Kegel collects postings from hardware, software, and applications news groups and offers links to many other ISDN-related sites. The ~ is the tilde on your keyboard. While still online in early 2000, the content seems to have been revised last in 1996.

Internet Engineering Task Force

The dominant form of publication from the IETF is the Request For Comment (RFC). There is a formal procedure to start them, but it doesn't slow the outpouring. There seems to be another one every few days. RFCs are numbered in sequence, as they are accepted for distribution. There are over 2300 now. They are available free on the Internet. Many ISP, college, and vendor sites have a recent index with titles.

Other

ISDN Ordering Codes; NIST/Bellcore/NIUF Seminar, June 1995. The National ISDN Users Forum held its last meeting in 1999, when it declared victory (ISDN had matured, was selling well) and transferred its document archive to NIST. Still plenty of useful information at www.niuf.nist.gov, particularly on National ISDN 2.

IEEE Communications Magazine, July 1990, special issue on Signaling System 7. Tutorials; reports from several countries.

J on Selected Areas in Communications, April 1994; issue on Common Channel Signaling Networks. Excellent overview of SS7, but otherwise contains scholarly analysis of networks functions.

The following data sheets on ISDN chip sets describe the detailed operation of the chips and thus illuminate and expand upon the rather dry text of the standards documents. Ask for the current edition.

T7250C *Enhanced User-Network Interface for ISDN and Proprietary Endpoints,* AT&T Microelectronics, August 1993. BRI chip for the S/T point of the BRI.

T7256 *Single-Chip NT1 (SCNT1) Transceiver,* AT&T Microelectronics, January 1995. U-point chip for the BRI.

Advance Information, MC145574 ISDN S/T Interface Transceiver, Motorola, August 1995.

Advance Information, MC145572 ISDN U-Interface Transceiver, Motorola, September 1995.

Index

ISDN A Practical Guide

ISDN A Practical Guide